Climate change and health in the Western Pacific Region

Synthesis of evidence, profiles of selected countries and policy direction

WHO Library Cataloguing-in-Publication Data

Climate change and health in the Western Pacific Region: synthesis of evidence, profiles of selected countries and policy direction

1. Climate change. 2. Environmental health. I. World Health Regional Office for the Western Pacific.

ISBN 978 92 9061 737 2 (NLM Classification: WB 700)

Photo credits

Cover photo: © WHO/Yoshi Shimizu

Inside pages: © WHO/Yoshi Shimizu: pp. xiv, 43 – © AFP: pp. 12, 38, 49, 62, 76, 84, 102 – © Photoshare: p. 71

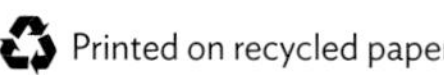

Contents

Figures

Tables

Preface

Thirty years from now the world is going to look very different, and exactly how it looks will depend on the actions we take today. Over the past 50 years, human activities – most significantly the burning of fossil fuels – have released increasing quantities of carbon dioxide and other greenhouse gases that trap heat in the lower atmosphere, thus accelerating the rate of global warming. Sea levels are rising, glaciers are melting, and extreme weather events are becoming more frequent and severe.

The health impact of climate change – for example, by expanding the range of mosquitoes that are vectors for dengue and malaria – is a complex issue that affects nations rich and poor.

Scientific evidence clearly shows the negative impact of climate change both on the planet and on living creatures. The World Health Organization estimated that in 2004, climate change was responsible for 140 000 deaths. Conservative estimates suggest that climate change will cause some 250 000 additional deaths per year before the middle of this century. The poorest and most vulnerable populations in low-income countries, particularly children and older people, are among those most at risk.

The Western Pacific Region, which includes many low-lying Pacific island countries and areas, is especially vulnerable to climate change, and the impact is not limited to climate-sensitive diseases. Changes in climate are expected to affect a wide range of environmental and social determinants of health, with heatwaves, rising sea levels and other extreme weather events contributing to a series of challenges ranging from food security to a scarcity of drinking-water and increases in communicable and respiratory diseases.

This report synthesizes information and approaches on climate change and health pertinent to Member States in the Western Pacific Region. It also examines efforts and initiatives by various experts and stakeholders, with an in-depth look at experiences in seven Member States that reflect the diversity of the Region. Finally, it offers recommendations for policy-makers.

There is serious concern about the impact of the changing climate. The WHO Regional Office for the Western Pacific has taken the initiative in addressing health issues related to climate change, but further action is needed to support efforts to confront climate change in Member States and in various sectors. Health must be mainstreamed into efforts to address climate change, and action must be coordinated and integrated across national boundaries and in all sectors.

The challenges are clear. Member States and stakeholders in communities across our vast Region must now work together if we are to mitigate and manage the health impacts of climate change.

Shin Young-soo, MD, Ph.D.
Regional Director for the Western Pacific
World Health Organization

Acknowledgements

Authors

Joshua Nealon, Hae-Kwan Cheong, Lachlan McIver, Nasir Hassan, Rokho Kim, Kristie Ebi, Michaela V. Pfeiffer, Kathryn Bowen, Jungsub Yeom, Hyenan Park, Suhyoon Choi and Yadav Prasad Joshi.

Reviewers

Marco Silvestri, Diarmid Campbell-Lendrum and Maria Elena Villalobos Pratskindly reviewed the document and made suggestions for its improvement.

Contributors

The authors would like to thank those people who provided assistance and advice throughout development of this report. Former and current WHO staff provided historical perspective, as well as support with technical issues, vulnerability analyses and adaptation plans. These contributors include Hisashi Ogawa, Hyenmi Chung, John Ehrenberg, Susan Mercado, Oyuntogos Lkhasuren, Bonifacio Magtibay, Tuan Nghia Ton, Chitsavang Chanthavisouk and Steve Iddings.

Arief Anshory Yusuf and Herminia Francisco performed regional vulnerability analyses. We would like to express our deep appreciation to Bettina Menne and her team in WHO European Centre for Environment and Health at the WHO Regional Office for Europe for providing a conceptual framework and insight on policy development through their pioneering work with Member States in the WHO European Region.

The authors also appreciate the network of researchers on climate change and health in Western Pacific Region who participated as consultants in the process of developing national action plans and organized a series of workshops to develop a framework for data analysis for an epidemiologic analysis of the health outcome data. The authors wish to specifically thank Yasushi Honda, Ho Kim and Masahiro Hashizume, as well as their students Jin-Hee Eum, Jin-Seob Kim, Yoon-hee Kim, Clara Tammy Kim and Chisato Chris Imai.

Other national and international experts also contributed to or conducted the vulnerability assessments and adaptation plans summarized in this report; without their diligent work the report would not have been possible. We would like to recognize in particular the work of Professor Jae-won Park, who passed away while on a mission conducting the vulnerability and adaptation assessment in the Lao People's Democratic Republic.

We would also like to recognize and thank ministries of health and other ministries and government bodies of Member States that have contributed directly and indirectly to the report through their hard work over many years.

This report was made possible by the generous contributions of donors, especially the Ministry of Environment, Republic of Korea. The authors also wish to acknowledge the contribution of the Government of Japan.

Abbreviations

AR5	5th Assessment Report
ASEAN	Association of Southeast Asian Nations
BAU	business as usual
CCAI	Climate Change and Adaptation Initiative of the Mekong River Commission
CCC	Climate Change Commission
CCD	Climate Change Department
CCO	Climate Change Office
CCSPH	Climate Change Strategy for Public Health
CO_2	carbon dioxide
CO_2e	carbon dioxide equivalent
COP	Conference of the Parties
CRED	Centre for Research on the Epidemiology of Disasters
CRHS	climate-resilient health system
DALYs	disability-adjusted life years
DANIDA	Danish International Development Agency
DPSEEA	Driving force-Pressure-State-Exposure-Effect-Action
ECPs	extended concentration pathways
EM-DAT	Emergency Events Database
ENSO	El Niño Southern Oscillation
FAO	Food and Agriculture Organization of the United Nations
GDP	gross domestic product
GHG	greenhouse gas
GIS	geographical information system
HAE	Health and the Environment unit, WHO Regional Office for the Western Pacific
HFCs	hydrofluorocarbons
HIA	health impact assessment
HSP2	Second Health Sector Strategic Plan
IOD	Indian Ocean Dipole
IPCC	Intergovernmental Panel on Climate Change

ISGE	International Support Group on Natural Resources and Environment
KOICA	Korean International Cooperation Agency
LDCs	least developed countries
LUCF	land use change and forestry
LULUCF	land use, land use change and forestry
MDGs	Millennium Development Goals
MMR	maternal mortality ratio
MONRE	Ministry of Natural Resources and Environment, Viet Nam
NAPs	national adaptation plans
NAPAs	national adaptation programmes of action
NCC	National Climate Committee
NCCAP	National Climate Change Action Plan
NCCC	National Climate Change Committee
NCDs	noncommunicable diseases
NFSCC	National Framework Strategy on Climate Change
NOAA	National Oceanic and Atmospheric Administration
NSAP	National Strategy and Action Plan
NSCCP	National Strategic Climate Change Plan
OCCD	Office of Climate Change and Development
OECD	Organisation for Economic Co-operation and Development
PHAP	Public Health Action Plan
PHE	Department of Public Health, Environmental and Social Determinants of Health, WHO
PFCs	perfluorocarbons
PM	particulate matter
ppb	parts per billion
ppm	parts per million
RCPs	representative concentration pathways
SPEED	Surveillance in Post-Extreme Emergencies and Disasters (Philippines)
SRES	Special Report on Emission Scenarios
SREX	Special Report on Managing the Risks of Extreme Events and Disasters to Advance Climate Change Adaptation
TWG	thematic working group
UN	United Nations
UNDP	United Nations Development Programme

UNEP	United Nations Environment Programme
UNESCAP	United Nations Economic and Social Commission for Asia and the Pacific
UNFCCC	United Nations Framework Convention on Climate Change
USAID	United States Agency for International Development
VIHEMA	Viet Nam Health Environment Management Agency
WHO	World Health Organization
WMO	World Meteorological Organization

Glossary of terms

Adaptation	Adjustment in natural or human systems to a new or changing environment. Adaptation to climate change refers to adjustment in response to actual or expected climatic stimuli or their effects, which moderates harm or exploits beneficial opportunities. Various types of adaptation can be distinguished, including anticipatory and reactive adaptation, public and private adaptation, and autonomous and planned adaptation.
Adaptive capacity	The ability of a system to adjust to climate change, including climate variability and extremes, to moderate potential damages, to take advantage of opportunities, or to cope with the consequences.
Biodiversity	The numbers and relative abundances of different genes (genetic diversity), species and ecosystems (communities) in a particular area.
Climate change	Climate change refers to any change in climate over time, whether due to natural variability or as a result of human activity.
Co-benefit	A climate change adaptation or mitigation strategy that has additional positive effects on health or other areas, e.g. reducing air pollution.
Coping capacity	The means by which people or organizations use available resources and abilities to face adverse consequences that could lead to a disaster. In general, this involves managing resources, both in normal times as well as during crises or adverse conditions. The strengthening of coping capacities usually builds resilience to withstand the effects of natural and human-induced hazards.
Exposure	The process by which an individual, community or ecosystem is affected by contact with a particular object, event or phenomenon – in this case, the effects of climate change.
Extreme weather event	An event that is rare within its statistical reference distribution at a particular place. By definition, the characteristics of what is called "extreme weather" may vary from place to place. An "extreme climate event" is an average of a number of weather events over a certain period of time, an average that is itself extreme, e.g. rainfall over a season.
Hazard	The capacity of an agent to produce a particular type of adverse health or environmental effect.
Health impact assessment	A systematic process to assess the actual or potential – and direct or indirect – effects on the health of individuals, groups or communities arising from policies, objectives, programmes, plans or activities.
Health risk assessment	The process of estimating the potential impact of a chemical, biological, physical or social agent on a specified human population system under a specific set of conditions and for a certain time frame.
Mitigation	The process of reducing the impact of climate change by reducing the driving forces thereof , i.e. reducing greenhouse gas emissions.

Resilience A health system that is capable to anticipate, respond to, cope with, recover from and adapt to climate-related shocks and stress, so as to bring sustained improvements in population health, despite an unstable climate.

Risk The probability that, in a certain time frame, an adverse outcome will occur in a person, group of people, plants, animals and/or the ecology of a specified area that is exposed to a particular dose or concentration of a hazardous agent, dependent upon both the level of toxicity of the agent and the level of exposure.

Sensitivity The degree to which a system may be affected, either adversely or beneficially, by climate-related stimuli. The effect may be direct, e.g. a change in crop yield in response to a change in the mean, range, or variability of temperature, or indirect, e.g. damages caused by an increase in the frequency of coastal flooding due to sea level rise.

Vector An organism, such as an insect, that transmits a pathogen from one host to another.

Vulnerability The degree to which a system is susceptible to, or unable to cope with, adverse effects of climate change, including climate variability and extremes. Vulnerability is a function of the character, magnitude, and rate of climate variation to which a system is exposed, its sensitivity and its adaptive capacity.

Ulaanbaatar, Mongolia

CHAPTER 1

Introduction

1.1 Background

> **"Climate change refers to a change in the state of the climate that can be identified...by changes in the mean and/or the variability of its properties, and that persists for an extended period, typically decades or longer."**
>
> Intergovernmental Panel on Climate Change (IPCC),
> Fifth Assessment Report, 2013

There is global consensus that a number of early effects of climate change have been observed over recent decades. These include but are not limited to increasing air and ocean temperatures, widespread melting of snow and ice, changing precipitation patterns, decreased frequency of cold days and nights, extended periods of drought, and an increase in the frequency of extreme weather events and their associated impacts, such as rising sea levels and deforestation. The impacts may be severe: tropical countries, for example, are at risk of devastating temperature increases, changes in precipitation patterns and increased heavy rainfalls from tropical cyclones. Sea level rises may have catastrophic impacts for low-lying Pacific nations (IPCC, 2013).

These climatic changes are anticipated to have a range of impacts on human health by both direct and indirect pathways. Most of these health impacts are anticipated to be unfavourable. The World Health Organization (WHO) estimated that climate change was causing over 140 000 excess deaths annually[1] by 2004 (WHO, 2009a), and the poorest populations in low-income countries, where vulnerability is highest, are likely to be disproportionately affected. Mechanisms of adverse health impacts include: the direct effects of heatwaves, cold spells and extreme weather events; impacts on mental health (Berry, Bowen & Kjellstrom, 2010); a lack of sufficient quantities and quality of freshwater; impaired nutrition due to compromised food security; increases in respiratory diseases associated with poor air quality; and increases in communicable disease incidence, both waterborne and vector-borne diseases. Health impacts, as with other effects of climate change, are likely to increase because even the most conservative climate projections indicate an escalation of climate change effects in the decades to come (Haines, Kovats & Campbell-Lendrum, 2006; Githeko, Lindsay, Confalonieri

1. This mortality estimate only covers four factors: malnutrition, malaria, floods and heatwaves.

& Patz, 2000; IPCC, 2014; McMichael, Friel, Nyong & Corvalan, 2008; McMichael, Woodruff & Hales, 2006; WHO, 2009a).

The health sector response to the threat of climate change has historically been modest, perhaps because of health professionals' typical requirement for proven causality between exposure and outcome, which is very difficult for a long-term phenomenon such as climate change where the long period of observation introduces inevitable confounding. Other factors, such as confusion within the health sector regarding tangible health adaptation strategies and a lack of access to adaptation funding mechanisms, may also play a role. The result has been a lack of health sector engagement. For example, in least developed countries (LDCs) and small island states, while 95% of national adaptation programmes of action (NAPAs) consider that climate change will impact health, only 3% of adaptation funding is for health (Manga, Bagayoko, Meredith & Neira, 2010).

Irrespective of these challenges and recognizing the opportunities of climate change adaptation funding, governments, partners and organizations have made commitments to respond and adapt to likely climate change health impacts. This was most clearly articulated by WHO Member States in 2008 at the World Health Assembly, which adopted resolution WHA61.19 on *Climate change and health*. The resolution urged Member States to take action on climate change, including the development and integration of health adaptation measures into existing plans (WHO, 2008a). The resolution specifically notes:

- the net global impact of climate change on human health is anticipated to be negative;
- vulnerable populations with the least ability to adapt will be most affected;
- climate change could jeopardize achievement of the Millennium Development Goals (MDGs);
- developing solutions to climate change impacts on health is a joint responsibility, and developed countries should assist developing countries in this regard;
- a priority in minimizing risk is the strengthening of health systems to enable them to respond to anticipated changes in public health needs;
- Member States should be consulted on the preparation of a global climate change and health work plan to scale up and address risks in a practical way.

A global climate change and health work plan was subsequently developed and amended by the WHO Executive Board at its 124th session in November 2008 (WHO, 2008b). The work plan has the specific aims: 1) to support health systems, in particular in low- and middle-income states and small island states, to enhance capacity for assessing and monitoring health vulnerability, risks and impacts due to climate change; 2) to identify strategies and actions to protect human health, particularly of the most vulnerable groups; and 3) to share knowledge and best practices.

The work plan has four distinct objectives:

Advocacy and awareness raising: to raise awareness that climate change is a fundamental threat to human health.

Engage in partnerships with other United Nations agencies and sectors: to coordinate with partner agencies within the United Nations (UN) system, and ensure that health is properly represented in the climate change agenda.

Promote and support the generation of scientific evidence: to coordinate reviews of the scientific evidence on the links between climate change and health, and develop a global research agenda.

Strengthen health systems to cope with health threats posed by climate change: to assist countries to assess their health vulnerabilities and build capacity to reduce health vulnerability to climate change.

For each objective, WHO committed to a number of actions, aiming to provide evidence and support capacity-building and implementation of projects to strengthen the health system response to climate change through activities at the country and regional levels and at WHO headquarters. WHO also works to ensure that health is appropriately considered in decisions made by other sectors such as energy and transport and provides the health sector voice within the overall UN response to climate change (WHO, 2008c).

1.2 Climate change, health and WHO response in the Western Pacific Region

The Western Pacific Region is comprised of 37 countries and areas and is home to more than one quarter of the world's population. The Region stretches over a vast area, from Mongolia to New Zealand, between 54° N and 53° S latitude, and from Central Asia to the Southwest Pacific, 73° E to 133° W longitude by land, and includes most of the surface area of the Pacific Ocean. Including the oceanic surface area, it includes more than one third of the global surface area. Its land mass embraces the eastern half of Eurasian land mass, the eastern part of South-East Asia, and the whole land mass of Oceania. It includes 22 Pacific island countries and areas, across the span of the Pacific Ocean. With unique geographies and population groups, the Region is exceptionally diverse, including LDCs, rapidly-emerging economies and developed nations. This diversity is reflected in climate change and health risks that vary considerably from one environment to another, from the vast Mongolian steppe and deserts to Asian megacities, the mountainous rain forests of Indochina and Papua New Guinea, and low-lying archipelagos and Pacific atolls.

Within the Health and the Environment (HAE) unit of the WHO Regional Office for the Western Pacific, climate change and health has remained a priority programme for over a decade. One of the earliest activities was a WHO Workshop on Climate Variability and Change and their Health Effects in Pacific Island Countries, held in Samoa in July 2000 (WHO, 2000). This workshop was jointly organized by a number of agencies, drawing conclusions that:

- climate is an important determinant of a number of health outcomes;
- there is increasing evidence of linkages between climate variability/change and health, and research is needed to strengthen these linkages;
- climate/health linkages are complex and must be viewed in the context of other environmental stressors and human activities;
- forecasting is an important tool for responding to climate/health risks and capacity should be developed in this area; and
- capacity-building at all levels is important to reduce vulnerability to climate variability and change.

Between 2000 and 2007, WHO issued a number of guidelines and documents, focusing on the most vulnerable areas which were often small island states. Training and other activities were conducted until the 2007 joint (Regional Office for South-East Asia and Regional Office for the Western Pacific) Workshop on Climate Change and Health in South-East and East Asian Countries, held in Kuala Lumpur, Malaysia. This meeting preceded the UN Climate Change Convention held in December 2007 in Bali, Indonesia.

Responding to climate change and health threats and following developments at the global level, the Regional Committee for the Western Pacific in September 2008 through resolution WPR/RC59.R7 endorsed the *Regional Framework for Action to Protect Human Health from the Effects of Climate Change in the Asia Pacific Region*[2] (WHO, 2008c). This is the earliest action taken on climate change among the six WHO regional offices. This resolution urges Member States:

- to develop national strategies and plans to incorporate current and projected climate change risks into health policies, plans and programmes to control climate-sensitive health risks and outcomes;
- to strengthen existing health infrastructure and human resources, as well as surveillance, early warning, and communication and response systems for climate-sensitive risks and diseases;
- to establish programmes to reduce greenhouse gas emissions by the health sector;
- to assess the health implications of the decisions made on climate change by other sectors, such as urban planning, transport, energy supply, food production and water resources, and advocate for decisions that provide opportunities for improving health;
- to facilitate the health sector to actively participate in the preparation of national communications and national adaptation programmes of action; and
- to actively participate in the preparation of a work plan for scaling up WHO's technical support to Member States for assessing and addressing the implications of climate change for health.

Prior to and following this resolution, Member States have shown considerable commitment to address associated health risks and have developed a number of strategies and plans for this purpose. These include health components of NAPAs developed by LDCs, other national climate change plans that include health components, and specific plans and initiatives targeting public health. Climate change adaptation activities initiated by other sectors – for example, targeting irrigation and agriculture, water and sanitation, coastal defence, or disaster risk reduction – will affect health through indirect pathways and should involve health sector inputs from an early stage.

Following endorsement of the *Regional Framework for Action to Protect Human Health from the Effects of Climate Change in the Asia Pacific Region,* HAE units at the Regional Office and country offices have been supporting Member States with health vulnerability assessments, development and implementation of national strategies and action plans, capacity-building in health adaptation to climate change, and the promotion of co-benefits, such as the reduction of greenhouse gas (GHGs) emissions for improved health. There has also been cooperation on climate change and health adaptation with WHO colleagues from other technical areas, particularly in the Malaria, other Vectorborne and Parasitic Diseases unit and the Emerging

2. The regional framework was developed jointly by WHO regional offices for South-East Asia and the Western Pacific during the Regional Workshop on Climate Change and Human Health in South-East Asia, 10–12 December 2007, Bali, Indonesia. The resolution presented in this report covers only the Western Pacific Region.

Disease Surveillance and Response unit at the Regional Office. Many of these activities are described in this report.

1.3 Projected impacts of greenhouse gas emissions on the global climate system

Understanding the health impacts of climate change and adapting to those impacts requires an understanding of the climate system, its variability, and likely changes in the global and regional climate systems. In contrast to weather, which is measured on a timescale of days to weeks and is continually changing, climate is the average state of the atmosphere in a given region over a long period. Climate change reflects a movement from this average and must be considered as a long-term – decades or longer – phenomenon showing a broad and consistent change in average weather and its extremes.

The earth's weather and climate are determined by complex interactions of the sun, the atmosphere, the oceans and the earth. The earth's land surface and oceans absorb a proportion of the solar energy that reaches the surface, reflecting the rest back to the atmosphere as infrared radiation. Some of this radiation is re-radiated back to the surface by atmospheric GHGs, including water vapour, carbon dioxide (CO_2), methane (CH_4), nitrous oxide (N_2O), perfluorocarbons (PFCs), hydrofluorocarbons (HFCs) and sulphur hexafluoride (SF_6). This re-radiation, known as the greenhouse effect, is what makes the planet habitable; without it, the average temperature of the planet would be approximately 33 °C colder. Burning of fossil fuels and deforestation increases the concentration of GHGs in the atmosphere, which increases the amount of solar energy re-radiated back to the surface of the earth (Fig. 1) with resulting increases in average ambient temperatures (IPCC, 2013).

Figure. 1. Illustration of the greenhouse effect

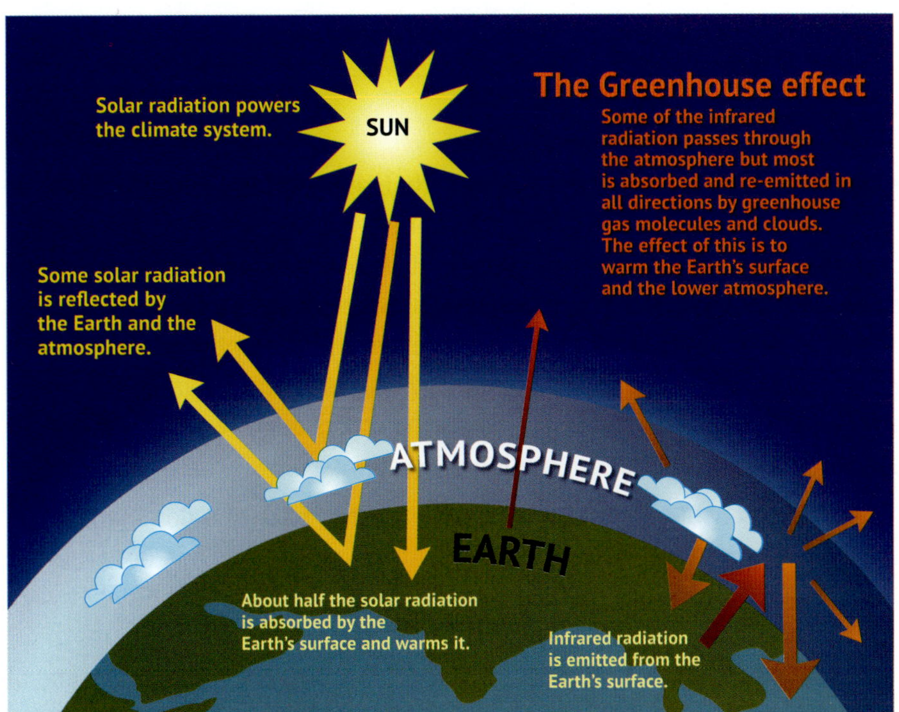

Source: IPCC, 2007b.

Water vapour is the most significant GHG, trapping more than 10 times the energy of the other GHGs. The most important human-derived (anthropogenic) GHG is carbon dioxide, whose concentration is increasing as a consequence of human activities. The atmospheric concentration increased from approximately 280 parts-per-million (ppm) in pre-industrial times to 400 ppm – a level first reached in May 2013. This exceeds by far concentrations for at least the past 800 000 years (Ahlenius, 2007; IPCC, 2013; National Oceanic and Atmospheric Administration [NOAA], 2013b). The primary source of this CO_2 is the combustion of fossil fuels such as oil, coal and gas, with a significant contribution from land use changes and the burning of plant matter. The ocean has absorbed about 30% of the carbon dioxide released from human activities, causing a measurable increase in the acidity of the oceans. The likely impacts of these emissions on global climate have been discussed for centuries, with the first attempt to quantify them in the late 19th century (Arrhenius, 1896). Additionally, the atmospheric concentrations of methane have more than doubled since pre-industrial times, the concentration of nitrous oxide has increased, and human activities have affected the concentrations of other GHGs (IPCC, 2013).

Impacts of global climate change are already occurring. In the 20th and 21st centuries, 12 separate years rank among the 14 warmest in recorded history: 2010 was the warmest year. Global average land and ocean temperature increased 0.85 °C between 1880 and 2013 (IPCC, 2013). Each of the last three decades has been successively warmer at the earth's surface than any preceding decade since 1850. The atmospheric water vapour content is also increasing because warmer air holds more moisture. Sea levels have risen over the past 50 years, partially as a consequence of the thermal expansion of water at warmer temperatures. Many other climatic changes have been observed and are reviewed elsewhere. Not only increasing ambient temperatures, but also changes in precipitation patterns are important for health. Over large areas of the earth, longer and more intense droughts have been observed since the 1970s (IPCC, 2013; NOAA National Climatic Data Center, 2012). The IPCC Fifth Assessment Report assessed the extent to which past changes in weather patterns could be attributed to climate change and the likelihood of their occurring in the future in the Asia-Pacific region. These assessments are summarized in Tables 1 and 2.

1.2.1 Climate projections

The Fifth Assessment Report (AR5) of the IPCC finds that rising temperatures worldwide are now unequivocal. Since the 1950s both the atmosphere and oceans have warmed, the amounts of snow and ice have diminished, and sea levels have risen (IPCC, 2013). A feature of AR5 is the emphasis on changes in the oceans: it is evident now that ocean warming dominates the increase in energy stored in the climate system, accounting for more than 90% of the energy accumulated between 1971 and 2010. This is of particular importance in the Pacific, where virtually all human settlement is on the coasts and is exposed directly to storm activity, sea level rise and changes in marine ecosystems.

AR5 concludes it is "extremely likely that human influence has been the dominant cause of the observed warming since the mid-20th century". At the same time, it is important to recognize that warming will not proceed uniformly, that there will be variability between years and from one decade to another, and the trends will differ also between regions (IPCC, 2013). Looking ahead, the IPCC judges that global surface temperature change for the end of the 21st century is likely to exceed 1.5 °C relative to 1850–1900 for most of the scenarios that have been explored (IPCC, 2013).

Table 1. Extreme weather and climate events: global-scale assessment of recently observed changes, human contribution to the changes and projected further changes for the early (2016–2035) and late (2081–2100) 21st century

Phenomenon and direction of trend	Assessment that changes occurred (typically since 1950 unless otherwise indicated)	Assessment of a human contribution to observed changes	Likelihood of further changes	
			Early 21st century	Late 21st century
Warmer and/or fewer cold days and nights over most land areas	Very likely	Very likely	Likely	Virtually certain
Warmer and/or more frequent hot days and nights over most land areas	Very likely	Very likely	Likely	Virtually certain
Warm spells/ heatwaves **Frequency and/ or duration increases over most land areas**	Medium confi- dence on a global scale Likely in large parts of Europe, Asia and Australia	Likely	Not formally assessed	Very likely
Heavy precipitation events **Increase in the frequency, intensity, and/or amount of heavy precipitation**	Likely more land areas with increases than decreases	Medium confidence	Likely over many land areas	Very likely over most of the mid-latitude land masses and over wet tropical regions
Increases in intensity and/or duration of drought	Low confidence on a global scale Likely changes in some regions	Low confidence	Low confidence	Likely (medium confidence) on a regional to global scale
Increases in intense tropical cyclone activity	Low confidence in long-term changes	Low confidence	Low confidence	More likely than not in the Western North Pacific and North Atlantic
Increased incidence and/or magnitude of extreme high sea level	Likely (since 1970)	Likely	Likely	Very likely

Note: projections are relative to the reference period of 1986–2005.

Source: modified from the Fifth Assessment Report, IPCC, 2013.

Table 2. Summary of Asia-Pacific projected regional changes in temperature and precipitation extremes

Region	Warm days	Cold days	Warm nights	Cold nights/ Frosts	Heatwaves/ Warm spells	Extreme precipitation	Dryness/ Drought
Asia (excluding South-East Asia)	High confidence, likely overall increase	High confidence, likely overall decrease	High confidence, likely overall increase	High confidence, likely overall increase	Medium confidence, spatially varying trends and insufficient data in some regions High confidence, likely more areas of increase than decreases	Low to medium confidence Low confidence due to insuf-ficient evidence or spatially varying trends. Medium confidence, increase in more regions than decreases	Low to medium confidence. Medium confidence, increase in eastern Asia
South-East Asia and Oceania	High confidence, likely overall increase	High confidence, likely overall decrease	High confidence, likely overall increase	High confidence, likely overall decrease	Low confidence (due lack of literature) to high confidence depending on region High confidence, likely overall increase in Australia	Low confidence (lack of litera-ture) to high confidence High confidence, likely decrease in southern Australia but index and season dependent	Low to medium confidence, inconsistent trends between studies in South-East Asia Overall increase in dryness in southern and eastern Australia High confidence, likely decrease northwest Australia

Note: regional observed changes in a range of climate indices since the middle of the 20th century.

Source: extracted and summarized from the Fifth Assessment Report, IPCC, 2013.

Additional drivers of climate variability include periodic oscillations that have global and regional impacts on weather. These systems can be useful in demonstrating the impacts of climate variability on human systems and health. The El Niño Southern Oscillation is a cycle in the ocean–atmosphere system in the Pacific that has important consequences for weather and climate around the globe: El Niño periods are characterized by unusually warm temperatures in the waters of the east-central equatorial Pacific, and the opposite, La Niña, by cool periods in the same waters. Global weather consequences include the weakening of easterly winds and increased rainfall in the eastern Pacific; flooding in the western United States and corresponding drought in Australia, Indonesia and Malaysia, sometimes associated with devastating fires. El Niño events occur irregularly at intervals of two to seven years and typically last 12–18 months. Due to effects in the global atmosphere, they also influence weather even in regions far from the tropical Pacific (NOAA, 2012). More specifically affecting the Western Pacific Region, an Indian Ocean Dipole (IOD) mediated by fluctuations in the temperature gradient between the western and eastern Indian Ocean has been described, with effects on wind, precipitation and ocean dynamics (Saji, Goswami, Vinayachandran & Yamagata, 1999). Effects include severe rainfall and drought, and the health impacts that have been described include cholera and malaria in South-East Asia and East Africa (Hashizume et al., 2011; Hashizume, Terao & Minakawa, 2009).

1.2.2 Regional climate change impacts and projections

A number of regional studies have been conducted, including as components of international studies, that aim to examine climate change impacts in the Asia-Pacific region. These studies contributed to IPCC findings that climate change has already had an impact on the region through mechanisms such as rising temperatures and increased frequency of extreme weather events and rainfall variability. Future impacts will include those on agriculture, marine and coastal environments, as well as biodiversity, with adverse impacts on human health. Importantly, climate change will challenge sustainable development by exacerbating existing pressure on natural resources and the environment (Cruz et al., 2007). Specific risks include

Box 1 RCP Scenarios

Representative concentration pathways (RCPs) are new scenarios that include time series of emissions and concentrations of the full suite of greenhouse gases and aerosols and chemically active gases, as well as land use/land cover (Moss et al., 2008). The word "representative" signifies that each RCP provides only one of many possible scenarios that would lead to the specific radiative forcing characteristics. The term "pathway" emphasizes that not only the long-term concentration levels are of interest, but also that there is a trajectory over time to reach that outcome (Moss et al., 2010).

RCPs usually refer to the portion of the concentration pathway extending up to 2100 for which integrated assessment models produced corresponding emission scenarios. Extended concentration pathways (ECPs) describe extensions of the RCPs from 2100 to 2500 that were calculated using simple rules generated by stakeholder consultations, and do not represent fully consistent scenarios.

Four RCPs produced from integrated assessment models were selected from the published literature and are used in the present IPCC assessment as a basis for the climate predictions and projections:

- **RCP 8.5:** One high pathway for which radiative forcing reaches over 8.5 W/m^2 by 2100 and continues to rise for some time (the corresponding ECP assuming constant emissions after 2100 and constant concentrations after 2250);
- **RCP 6.0** and **RCP 4.5**: Two intermediate stabilization pathways in which radiative forcing is stabilized at approximately 6 W/m^2 and 4.5 W/m^2 after 2100 (the corresponding ECPs assuming constant concentrations after 2150); and
- **RCP 2.6**: One pathway where radiative forcing peaks at approximately 3 W/m^2 before 2100 and then declines (the corresponding ECP assuming constant emissions after 2100).

Source: IPCC, 2013

decreased freshwater availability in Central, South, East and South-East Asia, particularly in large river basins; increased flooding in heavily polluted river delta regions; and rising health impacts of diarrhoeal diseases associated with floods and droughts (IPCC, 2014).

An additional and interesting regional climate change vulnerability mapping exercise was performed by the Economy and Environment Program for Southeast Asia. Vulnerability was derived from indices of exposure, sensitivity and adaptive capacity, providing an assessment of climate change vulnerability of countries at a provincial scale. While there are acknowledged limitations of the data-based approach, the method provides an objective means by which to compare countries at the subnational level. While not all countries in the Western Pacific Region were included, the resulting vulnerability map from this assessment indicates that most of Cambodia, the north-eastern section of the Lao People's Democratic Republic, the Mekong Delta region and the Philippines are among the most vulnerable areas in the Region to the impacts of climate change (Yusuf & Francisco, 2010).

1.4 Climate change and health synthesis report

A number of climate change and health activities have taken place in Member States of the Western Pacific Region. These include preparatory meetings and the development of plans, coordination of expert teams and their work supporting countries in the performance of climate change and health vulnerability analyses and adaptation plans, the submission of proposals for resource mobilization and fundraising, capacity-building training, and awareness-raising on climate change and health issues at all levels. A key feature of these activities has been the active participation of the non-health sector, including the agricultural, environmental, meteorological and other sectors, as appropriate. The cross-cutting nature of climate change initiatives facilitates such linkages and they represent a valuable entry point to other sectors that is often sought by health sector actors eager to access those responsible for health determinants that lie outside the health sector. Despite these activities and interactions, the concept of climate change and health remains somewhat elusive to health sector policy-makers. Implementation of adaptation plans has been limited or non-existent.

With limited time and resources, there is a need to coordinate the sharing of information, experience and knowledge among implementing countries and to reach a consensus on the way forward.

This report, therefore, was conceived to:

- provide a summary of evidence of climate change impacts on health in the Western Pacific Region;
- describe actions taken by countries, allowing others to learn from previous experiences and generate discussion about health sector adaptation in different geographical settings;
- stimulate rational resource mobilization to facilitate implementation of sound climate change and health adaptation plans; and
- provide policy direction for future action.

The country summary section of this report will focus on seven Member States in the Western Pacific Region: Cambodia, the Lao People's Democratic Republic, Mongolia, Papua New Guinea, the Philippines, the Republic of Korea and Viet Nam. These countries provide geographical, environmental and sociological diversity and expressed enthusiasm to share their experiences with others. As they are at different stages in the adaptation planning and implementing process, it is hoped their experiences will be of value to other countries. Papua New Guinea is included here because while it is located in the Pacific, it shares geographical and demographic similarities with the other Member States in the Western Pacific Region included in this report. A companion report for Pacific island countries, *Human health and climate change in Pacific island countries,* will describe the contrasting status of climate change and health in Pacific island countries and areas, and is being published in tandem. This report will be published separately, reflecting the potentially differing contexts and audiences, as well as many similarities.

The target audience of this report includes, but is not limited to:

- political leaders in the fields of climate change, health and the environment;
- policy-makers of Member States and associated organizations, development partners and their advisers;
- programme managers and technical staff responsible for IPCC and other technical submissions of countries;
- community leaders and members of civil society working within the fields of community empowerment, capacity-building and adaptation for climate resilience;
- the wider scientific community, including students and teachers at a range of educational institutions including universities;
- civil society, nongovernmental organizations and individuals with an interest in the area; and
- donor agencies, development banks and ministries of finance and foreign affairs with an interest in economic and social support for climate change adaptation within the health sector.

Banaue, Philippines

CHAPTER 2

The science of climate change and health

2.1 Background

> **"The issue now is not whether climate change is occurring, but how we can respond most effectively."**
>
> Dr Margaret Chan, Director-General of WHO,
> Cutting Carbon, Improving Health, 2009

Climate change transforms people's lives by extensively affecting ecosystems, agriculture, industry, air, water and the economy that support human health and well-being. There are many ways that climate change influences human health, such as direct impacts caused by heatwaves and other extreme weather events, and indirect impacts, such as outbreaks of infectious diseases and increases in allergic diseases, among other things.

When the ecosystem is transformed due to changes in climatic factors – such as patterns in temperature, precipitation, humidity and wind – the population size, habitat and encountered frequency of disease-spreading vectors also tend to change. A rise in temperature will push up the altitude of malaria- risk areas, hence expanding these areas. As can be seen in this example, previously disease-free zones may transform into risk areas due to climate change (Korea Centers for Disease Control and Prevention, 2010).

An important constraint in the field of climate change and health is the lack of empirical data associating exposure (climate change) to outcomes (health impact). The lack of robust statistical analysis is due primarily to the small magnitude of average climate change to date. For example, the global mean temperature increase between the 1951–1980 base period and 2012 is approximately 0.56 °C (Hansen et al., 2010). However, climate change does not act equally across broad averages, and microclimatic variations and shorter-term weather events may demonstrate the likely impacts of climate change on health and guide potential responses to them. In addition, many of the health effects are caused by indirect mechanisms that are poorly demonstrated using conventional epidemiological methods. In addition, the nature of local climate change and health impacts depends on local disease epidemiology and other health determinants, and this poses a challenge in attributing impacts. Other environmental or geographical variations may also provide demonstrations from which, in conjunction with laboratory data, inferences of future health impacts may be drawn. A number of studies and their findings are included in this chapter.

2.2 Greenhouse gas emissions in the Western Pacific Region

With the adoption of the *United Nations Framework Convention on Climate Change* (UNFCCC) in 1992 and the subsequent Kyoto Protocol at the third session of the Conference of the Parties (COP3), the reduction of GHG emissions is key to the mitigation of the global climate change. According to the UNFCCC, GHG emissions are rapidly increasing despite continued multilateral talks and activities. Table 3 shows that compared to 1971, GHG emissions had increased 2.2 times worldwide by 2010. Carbon dioxide is the major component of GHGs and globally, the Western Pacific Region accounts for 33.5% of total CO_2 emissions. Meanwhile, Asia including the Western Pacific and Middle East is the region with the highest rate of increase in CO_2 emissions since 1971. Compared to 1971, 1980, 1990 and 2000, CO_2 emissions in the Western Pacific Region have increased 5.3-, 3.6-, 2.5- and 1.8-fold, respectively. Four major countries are responsible for more than 90% of total CO_2 emissions in the Region (Table 3).

Table 3. Greenhouse gas emissions of countries in the Western Pacific Region, by year

Country	GHGs emission (million tonnes of CO_2 per year)					Proportion in WPRO (%)	Increase (%)			
	1971	1980	1990	2000	2010		vs 1971	vs 1980	vs 1990	vs 2000
Australia	144.1	208.0	260.0	338.8	383.5	3.8	2.7	1.8	1.5	1.1
Brunei Darussalam	0.4	2.6	3.4	4.6	8.2	0.1	20.6	3.1	2.4	1.8
Cambodia				2.0	3.8	0.0				1.9
China	800.4	1 405.3	2 211.3	3 037.3	7 217.1	71.1	9.0	5.1	3.3	2.4
Hong Kong (China)	9.2	14.5	32.8	39.8	41.5	0.4	4.5	2.9	1.3	1.0
Japan	758.8	880.7	1 064.4	1 184.0	1 143.1	11.3	1.5	1.3	1.1	1.0
Korea, Rep. of	52.1	124.4	229.3	437.7	563.1	5.5	10.8	4.5	2.5	1.3
Lao PDR			0.2	1.0	1.8	0.0			9.0	1.8
Malaysia	12.7	24.3	49.6	112.7	185.0	1.8	14.6	7.6	3.7	1.6
Mongolia			12.7	8.8	11.9	0.1			0.9	1.3
New Zealand	13.7	16.4	23.4	30.9	30.9	0.3	2.2	1.9	1.3	1.0
Philippines	23.0	33.3	38.2	67.5	76.4	0.8	3.3	2.3	2.0	1.1
Singapore	6.1	12.7	29.4	47.7	62.9	0.6	10.3	5.0	2.1	1.3
Viet Nam	16.1	14.8	17.2	44.0	130.5	1.3	8.1	8.8	7.6	3.0
Other countries	39.3	89.4	124.6	229.7	291.1	2.9	7.4	3.3	2.3	1.3
WPRO	1 876.0	2 826.4	4 096.3	5 585.6	10 148.8	100.0	5.4	3.6	2.5	1.8
World	14 064.8	18 042.2	20 973.9	23 509.1	30 276.1	298.3	2.2	1.7	1.4	1.3

Source: data originated from UNFCCC, 2013. Emissions of the Lao People's Democratic Republic cited from the United Nations Statistics Division.

Figure 2. Spider map of the selected countries showing proportion of greenhouse gas emissions by sector

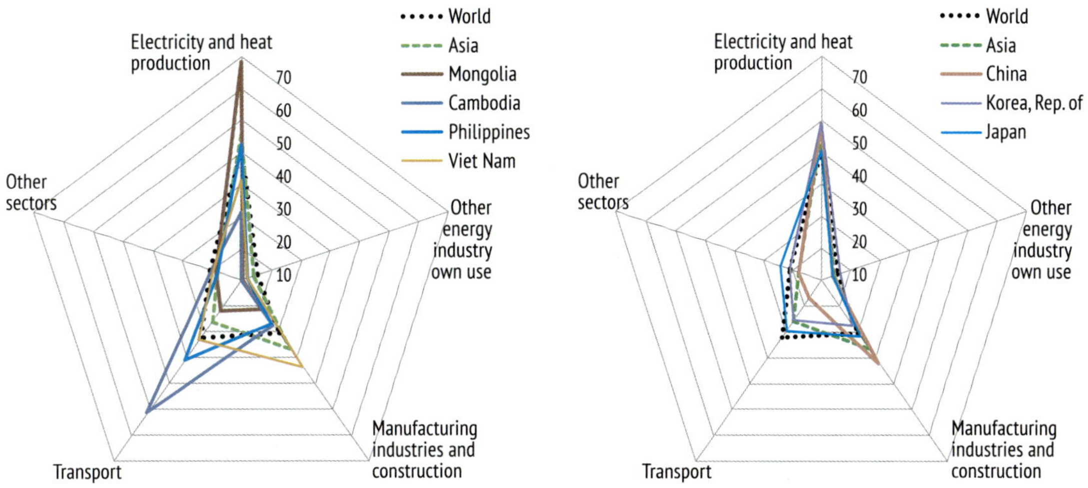

Source: drawn from data originated from UNFCCC, 2103; International Energy Agency, 2012.

While those countries most vulnerable to the impacts of climate change tend not to be significant contributors to emissions, the rate of CO_2 emissions in some developing countries is increasing at speed due to rapid economic development.

The major source of emissions varies depending on the level of economic development, industrial structure, climate and source of energy. Worldwide, energy and heat production are responsible for more than 50% of GHG emissions, followed by transport, manufacturing industries and other sectors. These ratios are broadly similar to those observed in Asia as a whole (Fig. 2).

2.3 Climatic determinants of human health

Human behaviour and survival are intimately associated with environmental determinants, and humans can not survive where essential environmental requisites such as water, sunlight and ambient temperatures conducive to survival are absent. Changes in weather and climate will have important impacts on human health via a number of direct and indirect mechanisms, some of which are illustrated in Figure 3. Recognizing the associations between health and climate, WHO and the World Meteorological Organization (WMO) have published the *Atlas of Health and Climate*, describing associations, pathways and outcomes of climatic events on health outcomes. The atlas also maps areas at risk and is notable for the breadth of its outcomes, covering infectious diseases, emergencies such as food shortages and drought, and emerging environmental challenges (WHO & WMO, 2012).

An important feature of the report is that it summarizes published literature that disproportionately arises from developed countries with established surveillance systems able to detect special changes in disease incidence and their determinants, and with the requisite research capacity. In contrast, the burden of the health impacts of climate change are overwhelmingly likely to be most keenly felt in developing countries where the adaptive capacities of populations are limited and health infrastructure is least mature.

Figure 3. **Framework of impact of climate change and variability on human health in the Western Pacific Region**

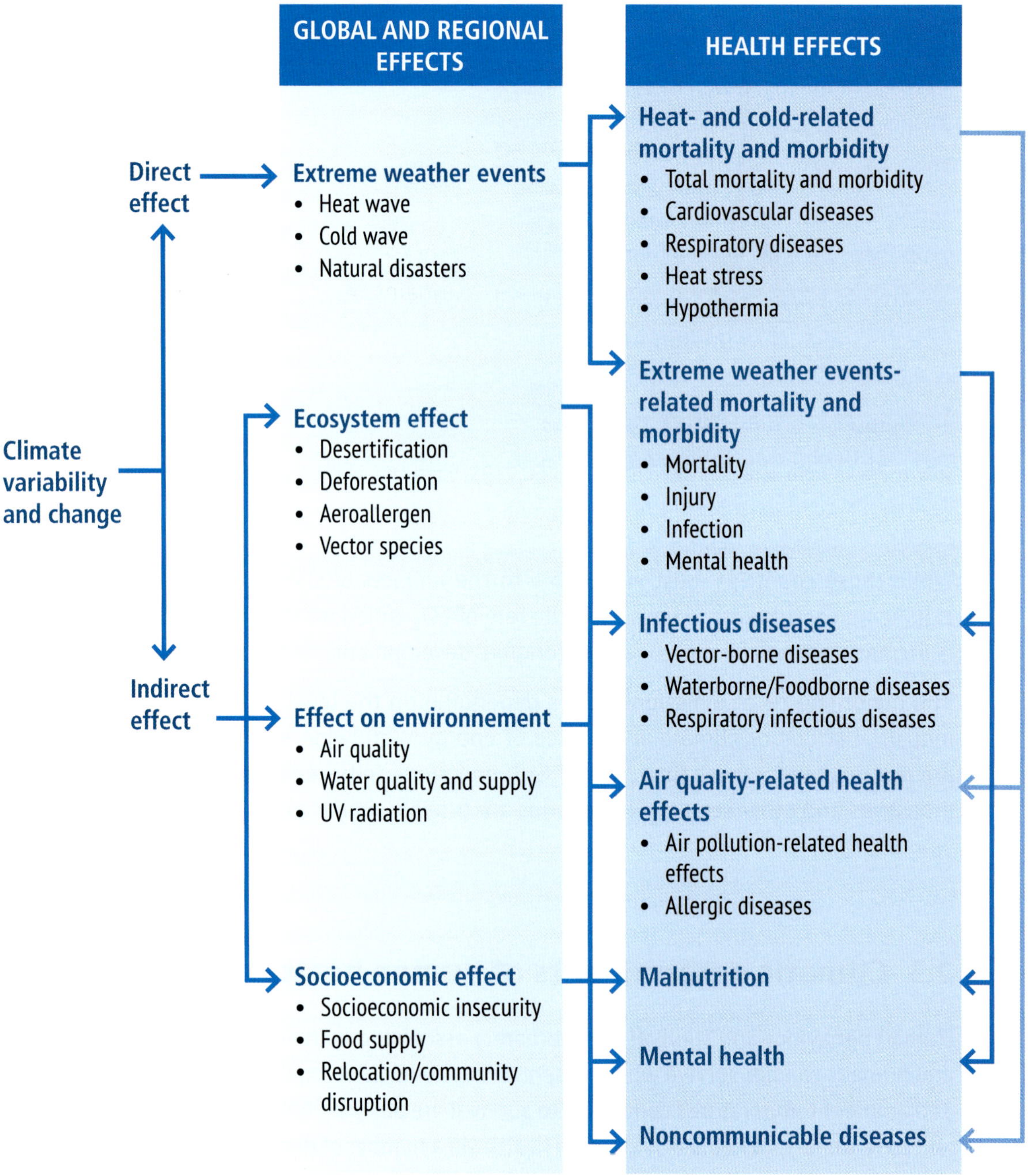

The report concludes that, to date, there is evidence that climate change has impacted the distribution of vectors of some infectious diseases, altered the distribution of allergenic pollen species and increased the number of heatwave-related fatalities. According to the published literature and climate projections, the future impacts of climate change on human health are likely to:

- increase the burden of diarrhoeal diseases;
- alter the range of the vectors of some communicable diseases;

- have mixed effects on malaria (contraction in some areas, expansion in others, with the overall balance projected to be expansion);
- increase the number of people at risk of infection with dengue;
- increase morbidity and mortality associated with extreme weather effects such as storms, floods, heatwaves, fires and droughts;
- increase cardio-respiratory mortality and morbidity associated with ground-level ozone; and
- increase malnutrition and associated disorders, including those related to child growth and development.

Despite these predictions and a broad range of peer-reviewed literature, health sector engagement in climate change adaptation remains minimal for a number of reasons. While studies have attempted to measure climate change impacts, these effects are felt in combination with a wide range of environmental, social, public health and developmental changes that interact and may reduce or magnify true impacts. The issue of attribution of observed changes has been addressed for some specific diseases, with some arguing that a number of criteria must be met before considering that epidemiological changes are genuinely climatically determined (Kovats et al., 2001) . These include:

- evidence of biological sensitivity to climate;
- meteorological evidence of climate change (in a specified area); and
- evidence of entomological/epidemiological change in association with climate change.

In cases where long-term surveillance data are available in tandem with historical meteorological data, it is possible that statistical methods could be used to investigate the climate change–health relationship. In many cases, particularly in developing countries, such analyses should be approached with caution to ensure comparability between historical data. This is especially pertinent given the relatively small magnitude of historical climate change: over long time periods, changes in epidemiological surveillance systems, case definitions, locations of meteorological stations and other confounders are likely to impact results (Haines et al., 2006).

Additionally, health impacts of climate change will be felt at the local level where the microclimate, the availability of infrastructure and services, and individual population characteristics and vulnerabilities will be locally specific. For this reason, global models may be of limited value in projecting local-level health impacts. The Western Pacific Region has a relatively small number of studies on climate change and health, therefore limiting the understanding of risks and preventing timely adaptation measures from being implemented.

The health risks of climate change are a function not just of the changing weather patterns occurring with climate change, but also the natural and human systems exposed to these changes, and the underlying vulnerability of the exposed systems. This means that understanding the risks requires considering not just climate change, but also current and future vulnerability and exposure.

2.4 Evidence of health impacts in the Western Pacific Region

Within the Western Pacific Region, evidence of climate change impacts has been limited. However, retrospective epidemiological and climate data are available from a number of countries, and exploratory studies have examined possible associations and historical trends that might inform future studies and provide an initial understanding to support policy development. These analyses tend to have been performed with the following assumptions:

- analyses are based on retrospective data and are not an attempt to ascertain future disease burdens;
- an understanding of regional climate phenomena and how they impact local weather and health would provide a useful, medium-term model of climate change;
- the relationship between climate and health is heavily dependent on the local determinants of health. It is therefore appropriate to use a local scale of analysis wherever possible.

In addition, a number of studies on climate change impacts on key health outcomes have been conducted. A selection of these studies are included below, where we describe and summarize the evidence of the impacts of climate change on health.

2.4.1 Changing trends in extreme weather events

One of the most prominent health risks of climate change is an increase in extreme weather events. Much of the Western Pacific Region consists of archipelagos and island nations; coastlines are long and countries have a long history of cyclones, floods, landslides, typhoons and other associated events. The Centre for Research on the Epidemiology of Disasters (CRED) maintains the International Disaster Database (EM-DAT), classified by disaster subgroup (biological, geophysical, hydrological and meteorological). This number has been increasing markedly since the 1950s, and part of the reason is more people moving into harm's way (Fig. 4) (CRED, 2013). The Asian region is particularly exposed (Tables 4 and 5). Climate models predict an increased frequency of extreme weather events that will have direct and indirect health impacts. An understanding of likely future trends would allow policy-makers to prepare for future impacts, strengthen protective infrastructure and early warning/response capacity, and provide information to vulnerable populations.

2.4.2 Heatwaves

Heatwave-related health effects are one of the direct effects of climate change. Heatwaves are related to increased mortality due to cardiovascular disease and respiratory disease, as well as general mortality. They are also associated with increased morbidity of cardiovascular disease and respiratory disease, which manifest as more visits to emergency departments or hospital admissions (Kovats et al., 2008). Thresholds for these effects to become apparent are lower in countries with higher latitude, and higher or obscure in the subtropical or tropical countries (Patz & Olson, 2006; Hajat & Kosatsky, 2010).

In Japan, an increase of ambient temperature in the summer of 2010 of 1.64 °C above the normal average summer temperature was accompanied by an outbreak of heatstroke nationwide (Kondo et al., 2011). Response to the increased air temperature was observed

Figure 4. **Time trend of disasters (climatological, hydrological and meteorological) from 1900–2012, Western Pacific Region countries and world**

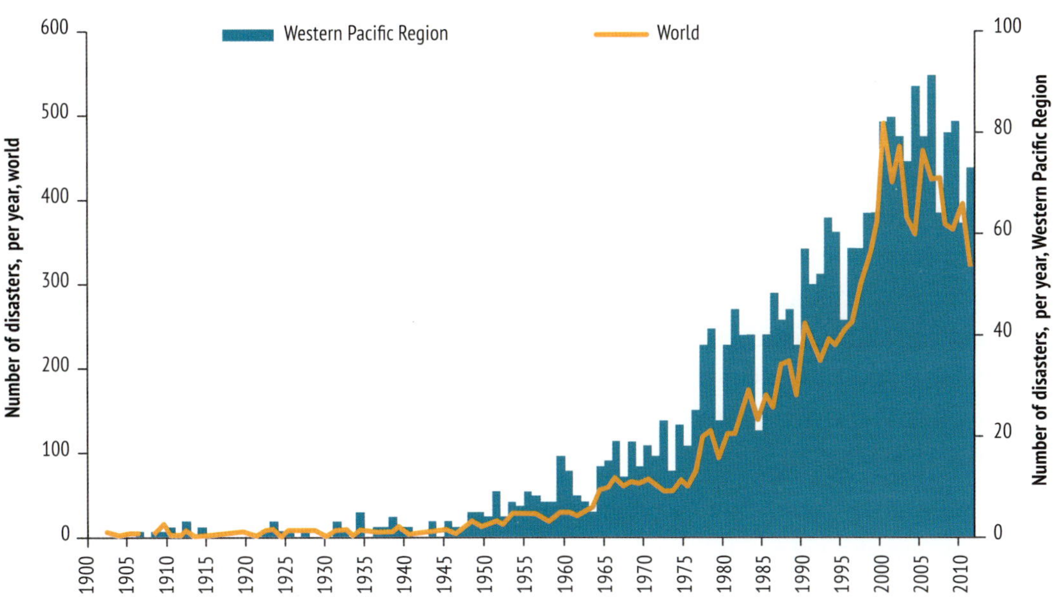

Source: drawn based on database of EM-DAT, CRED, 2013.

Table 4. **Average physical exposure to tropical cyclones and floods in different global regions**

	Africa	Asia	Australia, New Zealand	Central and South America	Islands (Indian Ocean, Pacific Ocean, Caribbean and other islands)	North America
1970	500	68 000	50	30	1 910	2 610
2030	2 280	125 950	100	100	3 490	4 870

Table 5. **Average physical exposure to floods assuming constant hazard**

	Africa	Asia	Australia, New Zealand	Caribbean	Central and South America	Europe	North America
1970	850	29 780	30	70	550	1 650	640
2030	3 640	77 640	60	180	1 320	1 870	1 190

Note: average physical exposure to tropical cyclones assuming constant hazard (in 1000s of people per year)..

Source: modified from IPCC, 2012, Handmer et al., 2012.

in most Japanese cities, but the magnitude of the increase depended on the vulnerability of the population.

Increasing temperature has a consistent dose–response relationship with mortality, which typically shows increasing mortality following either extremely hot or extremely cold temperatures (Honda et al., 2009; Hajat & Kosatsky, 2010). The lowest point of mortality is designated as an "optimum temperature" for the population in the specific area and subgroup. In general, this optimum temperature is higher in a hotter locale and lower in a cooler locale, reflecting the acclimatization of the population to the climate. Above the optimum point, mortality increases in a linear fashion, enabling the prediction of mortality from heatwaves in a specific area and population. This phenomenon is most prominent in countries of the temperate zone. The health impacts of heatwaves are observed at the beginning of the hot summer months, after which the population seems able to take appropriate protective measures. For this reason, early warnings must be prompt in order to prevent avoidable mortality (Ha & Kim, 2013).

This relationship and the existence of population-specific thresholds also provide important information on the vulnerability of population subgroups, and studies have identified subgroups vulnerable to the effects of heatwaves and mortality. These include those with biological or socioeconomic factors that render them more susceptible or less resilient to heat-related stress – typically older people, those with cardiovascular or respiratory disease, and those of lower socioeconomic status. Children are also susceptible to the effects of heat (Xu et al., 2012). With increasing temperature, more hospital admissions, respiratory diseases, fever episodes and diarrhoeal disorders have been reported, especially in children under five years old (Onozuka & Hashizume, 2011b).

There is a clear effect of high temperature on mortality above certain thresholds, and this effect is consistently observed across Asian cities. The primary cause of these fatalities is respiratory and cardiovascular events (Chung JY et al., 2009). Socioeconomic factors play a role in susceptibility: poorer people living in city centres, in suboptimal housing and with less cooling capacity, are generally located within urban "heat islands", creating hot spots of vulnerability during heatwaves (Kim et al., 2012b; Bambrick et al., 2011)

In parallel, exposure to cold is also a risk for mortality and morbidity including via cardiovascular disease. Cold sensitivity is affected by latitude: in colder places such as Yakutsk, Russia, temperature-dependent mortality increases are less prominent, with the exception of increased infant mortality and life expectancy (Young & Makinen, 2010). However in subtropical and temperate areas, higher mortality during colder weather has been observed (Revich & Shaposhnikov, 2010; Burkart et al., 2011; Yang et al., 2011).

In light of the many uncertainties, projections of the impact of climate change on population health should be presented as a range of plausible outcomes. The *Garnaut Climate Change Review* estimated that annual temperature-related deaths (winter and summer) in Australia for unmitigated climate change would increase by 1250 deaths in 2070, and 8628 deaths in 2100, compared with no climate change (Bi et al., 2011).

2.4.3 Increased incidence of communicable diseases including vector-borne

• Mosquito-borne diseases

One of the most prominent health risks of climate change, often cited in the international scientific literature, is an increase in the geographic range, seasonality and incidence of vector-borne diseases, due predominantly to the possible expansion in the range of arthropod vectors from changing weather patterns. Laboratory studies, models and limited observational data from the field indicate that such changes already may have taken place.

1. Malaria

Malaria causes a high disease burden in endemic countries. Globally, overall incidence of malaria is declining, and the number of deaths from malaria has decreased from 985 000 in 2000 to 781 000 in 2009 (*WHO World Malaria Report,* 2012c), thanks to extensive antimalarial activities. Over 90% of cases are in Africa, but several Western Pacific Region countries remain affected. The disease has long been discussed in the context of climate change because many of the determinants of transmission, including vector populations and behaviour, parasite incubation time and associated human behavior, are climatically mediated (Martens et al., 1995; Githeko et al., 2000).

Within the Western Pacific Region, Papua New Guinea is the country with the greatest malaria burden, accounting for 82.0% of all cases developed in the Region. In all other countries, malaria incidence has fallen markedly since 2000, although the decreasing tendency is only slight in the Philippines (Fig. 5).

An interesting feature of the disease in countries of the Western Pacific Region is the persistence of *vivax* malaria in China and the Republic of Korea, both temperate countries. Although there is a clear decline of incidence, recent reports show that *vivax* malaria is highly dependent on climate factors, with higher risks of outbreaks in a warming climate (Kim et al., 2012a).

Figure 5. Number of cases of malaria in the Western Pacific Region between 2000 and 2009

Left: malaria cases in selected Western Pacific Region countries
Right: time trend of malaria case in four countries of project

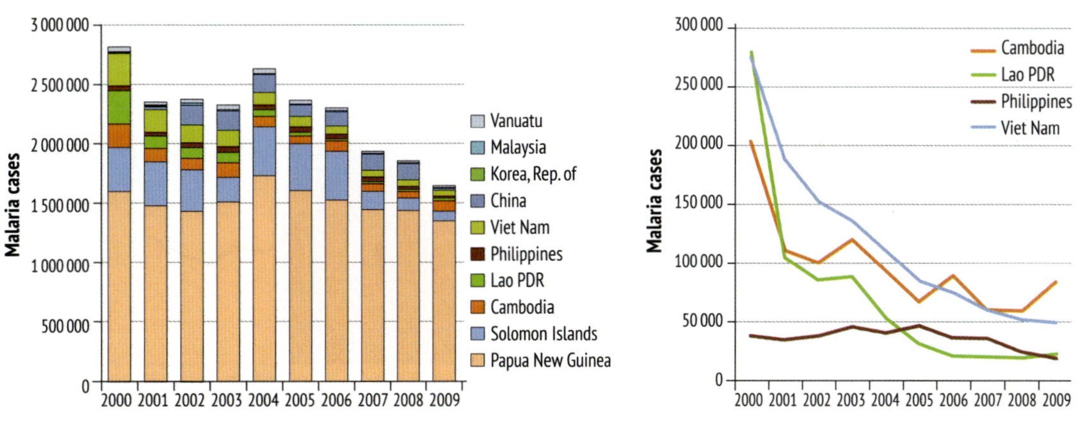

Source: based on WHO, 2012c.

Figure 6. Time trend of malaria incidence in four geographic regions of Papua New Guinea, 1997–2009

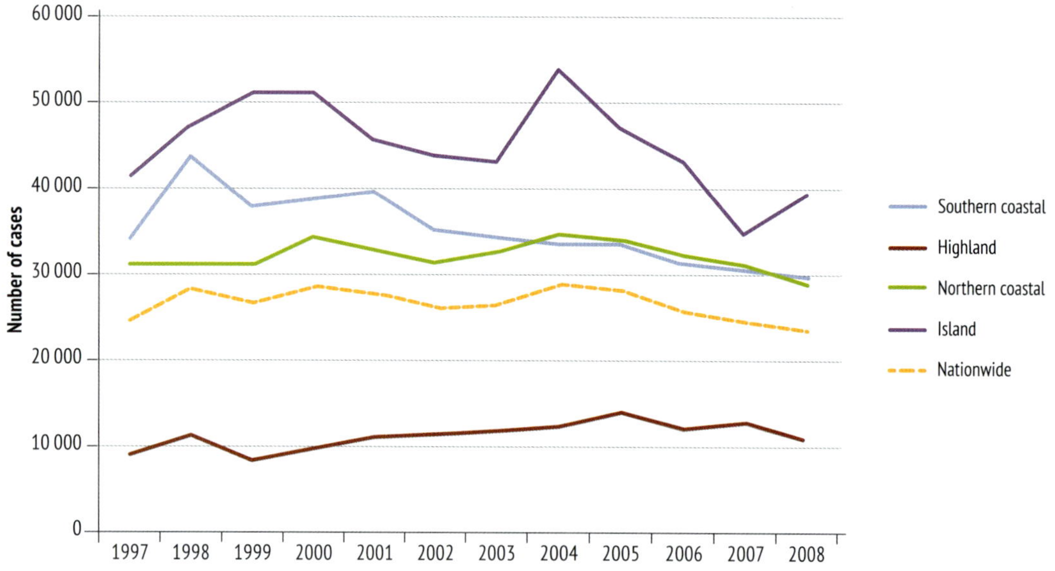

Source: National Department of Health, Papua New Guinea, 2011.

Figure 7. Trend of rainfall and temperatures over last decade in Papua New Guinea

a) Daru, southern coastal region – **b)** Goroka, highland region
c) Madang – **d)** Wewak, in northern coastal region

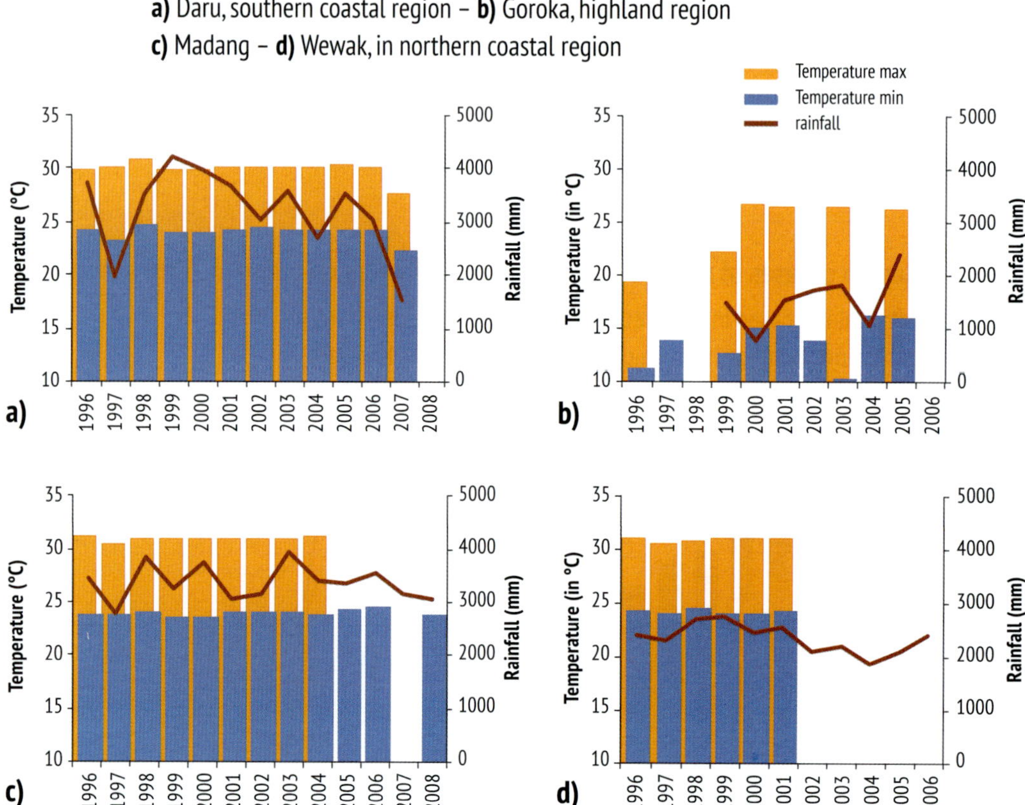

Source: National Department of Health, Papua New Guinea, 2011.

Incidence trends of malaria in Papua New Guinea may provide an insight into the impact of climate change on the disease elsewhere. Nationwide, the incidence of malaria is decreasing in most coastal and island areas, but in highland areas incidence is increasing, perhaps as a consequence of increasing rainfall and temperatures in the highlands over the past 12 years (Fig. 6 and Fig. 7).

In highland regions, the trend of malaria incidence differed by altitude. In the altitude range below 1500 m, incidence was highest and the rate of increase was greatest. An increase in altitude was associated with decreasing incidence to an altitude of 1700 m. Above 1700 m, incidence was lowest and incidence was stationary. Since 1996, a gradual increase in incidence at lower altitudes can be observed, but this has not yet extended to higher altitudes in which environmental conditions do not support transmission (Fig. 8). This could change as the climate warms.

In the temperate zone, malaria has a typical seasonality with a highest peak during the summer (Fig. 9). Because of a long winter and lack of an adequate vector mosquito, transmission of malaria is mostly limited to *vivax* malaria, which has evolved to adjust to the temperate zone and survive the long winter, which is not favourable for transmission. Therefore, the transmission period is limited to the summer season and vector dependencies on climate factors are closer. Analysis of climate factors in the mid-latitude zone of East Asia shows that malaria risk increases in relation to higher temperature and rainfall, with a time interval to allow for the vector life-cycle and incubation period (Kim et al., 2012a). This time relationship between climate factors and malaria development suggests the effectiveness of early warnings for malaria.

Figure 8. Incidence of malaria in highland Papua New Guinea according to altitude range, Eastern Highland Province, 1997–2009

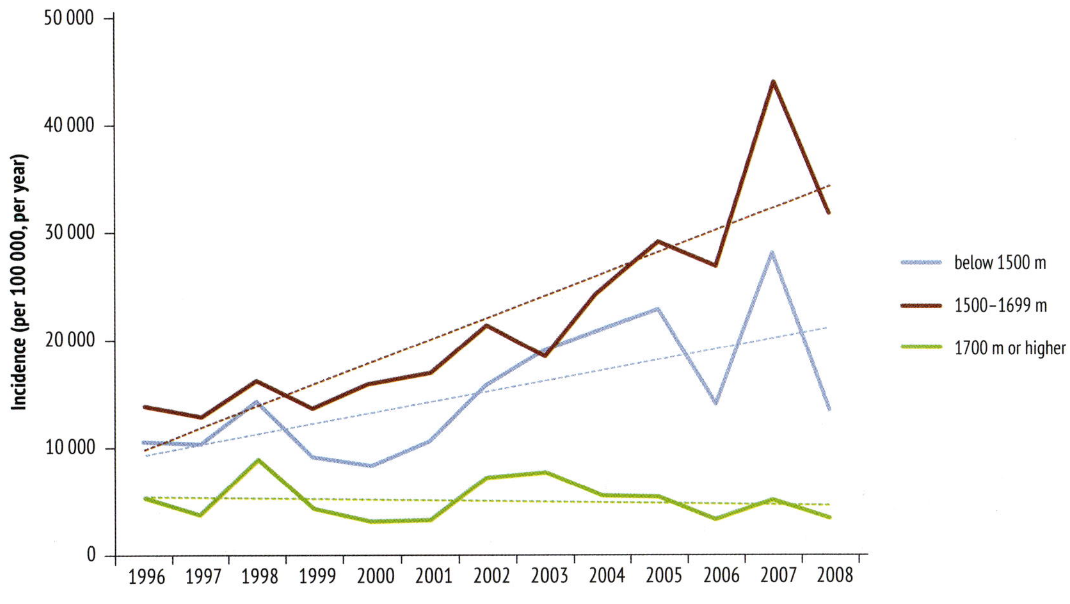

Note: dotted line denotes trend line for incidence in each altitude range.

Source: National Department of Health, Papua New Guinea, 2011.

Figure 9. Seasonality of malaria in Cambodia (left), **Republic of Korea** (middle), **and Papua New Guinea** (right)

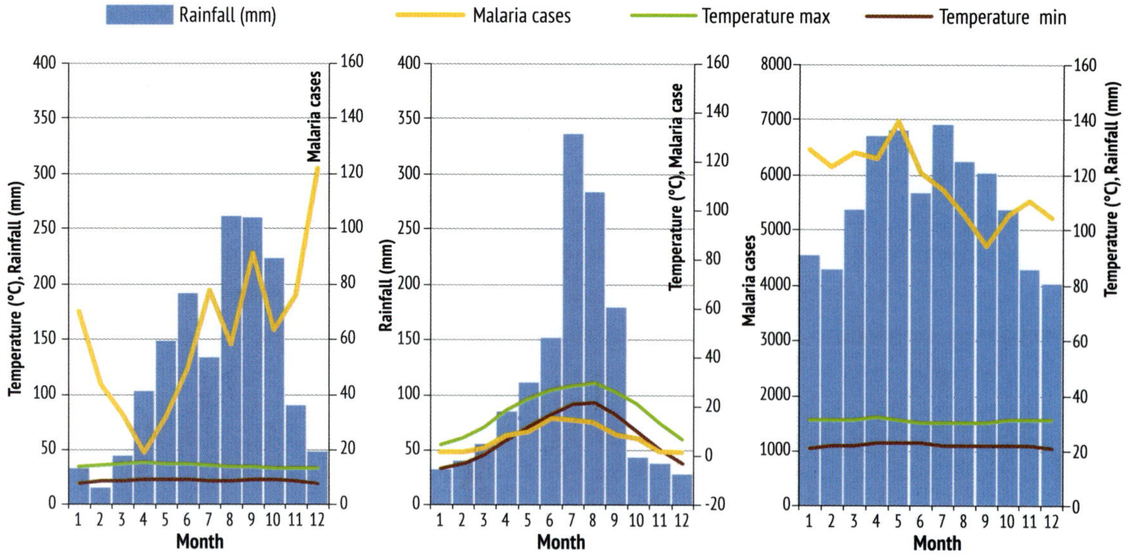

Rainfall (mm) — Malaria cases — Temperature max — Temperature min

Source: WHO Regional Office for the Western Pacific, 2012b.

2. Dengue

Dengue is transmitted by *Aedes* species mosquitoes, predominantly *Aedes aegypti*, although *Aedes albopictus* is a secondary vector in some settings. The distribution of the vector is determined climatically: cases occur only in tropical and subtropical areas located within the January and July 10 °C isotherms (Fig. 10) (WHO, 2012a).

Figure 10. Countries or areas at risk of dengue, 2011

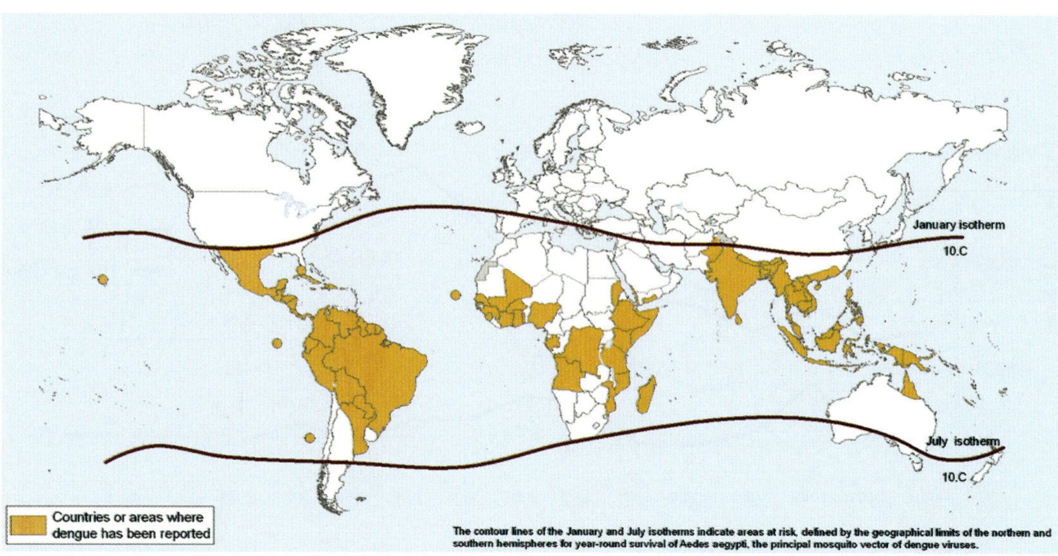

Source: WHO, 2012a.

Globally, the dengue burden is increasing, and – in combination with factors such as increased urbanization and dispersal of viruses and vectors – the presence of favourable climatic conditions is one reason for this expansion. The relationship between climate and disease incidence is clearly shown in Cambodia, where dengue outbreaks occur annually. The magnitude of outbreaks varies from year to year according to a number of intrinsic and extrinsic determinants, but cases are extremely seasonal. Monthly incidence is highly associated with monthly rainfall and temperature (Fig. 11). The underlying biological mechanism responsible for this seasonal fluctuation is presumed to be an increase in *Aedes* vector mosquito density during the wet season as breeding sites become more abundant, but this has not been conclusively demonstrated.

A number of studies have been conducted in the Western Pacific Region assessing the likely impacts of climate change on dengue distribution and incidence. Time-series studies in Viet Nam, for example, conclude that higher dengue incidence is associated with higher rainfall, humidity and temperatures, and that the dengue burden will increase with climate change (Pham et al., 2011). In Singapore, dengue cases can be predicted by high maximum and minimum temperatures (Pinto et al., 2011). Interestingly, an Australian model predicts climate change will result in an increased potential range of *Aedes aegypti* with accompanying dengue outbreak risks. However, this increase is as a result of an increased number of water storage containers, rather than via direct climatic impacts on vector biology (Beebe et al., 2009).

Incidence of dengue has increased more than 30-fold over the past 50 years. The evidence linking the rapid increase of dengue and climate change is not conclusive: review studies in the Asia-Pacific region indicate that while climatic changes are likely to impact the seasonal and geographic distribution of dengue through modification of climatically dependent biological pathways, no clear evidence exists that such a change has occurred. Dengue epidemiology is closely associated with human behaviour, and sociological variables should be included

Figure 11. Relationship between monthly dengue reports (red line) **and monthly rainfall** (blue bars) **in Siem Reap and Phnom Penh, Cambodia***

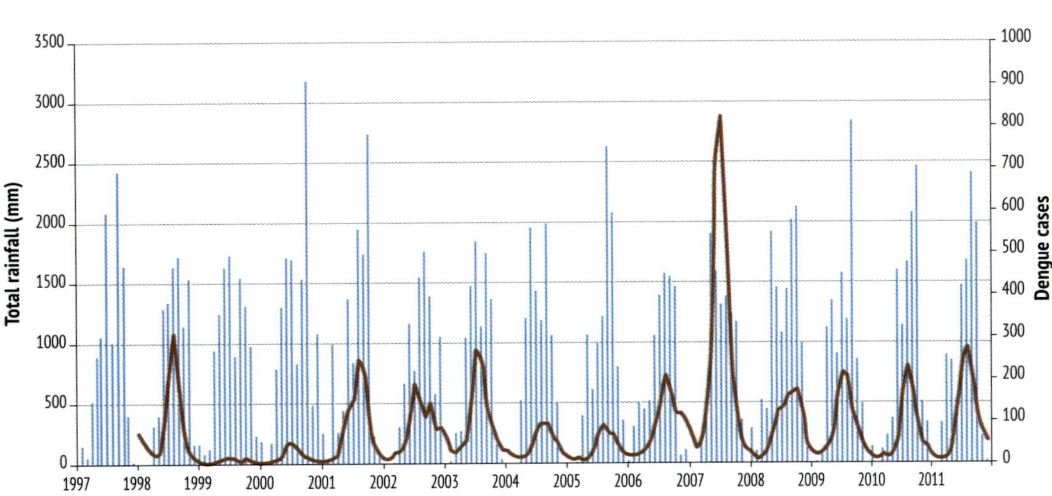

* Data supplied by Ministry of Health and Ministry of Water Resources and Meteorology, Kingdom of Cambodia.

Source: WHO and WMO, 2012.

in future analyses (Banu et al., 2011). In China, a review of climate change impacts on mosquito-borne diseases (dengue, Japanese encephalitis and malaria) found that evidence was inconclusive and geographically inconsistent (Bai et al., 2013).

It is important to note that although there is limited ability to attribute changes in the geographic range, seasonality or incidence of vector-borne diseases to climate change, this does not mean it will not be possible to do so in the future. It can be expected that long-term increases of winter temperatures in higher latitude regions may facilitate the expansion of *Aedes* vectors both north and south which, combined with higher disease burdens in the tropical and subtropical areas and increased tourism and travel, may result in an expansion of the transmission zone. This was demonstrated in the 2012 dengue outbreak in Madeira, Portugal, with over 2000 locally acquired cases. This was the first outbreak in Europe since the 1920s.

- **Tick-borne diseases**

Within the Western Pacific Region, the highest profile vector-borne diseases tend to be those transmitted by mosquitoes, but diseases transmitted by ticks and fleas are also a public health concern. Ticks are excellent vectors of zoonotic diseases because immature ticks feed on other animals before seeking a larger host, such as a human. There is increasing evidence from northern countries that the range of tick-borne diseases is increasing as a consequence of climate change, and particularly due to a reduction in the number of very cold spells (Jaenson & Lindgren, 2011). While these relationships are still unclear, some studies have taken place in Mongolia examining tick host-seeking behaviour and relationships with temperature and other environmental parameters, with a view to understanding likely climate change impacts. Initial findings indicate that ticks are extremely sensitive to local weather conditions: they are most active when temperatures are 6 °C to 10 °C and when humidity is low (WHO, 2012b).

In Mongolia, plague is transmitted primarily to young men and boys who become infected after exposure to infected fleas living on marmot or ground squirrel hosts handled as a consequence of hunting or playful trapping of the animals. Incidence has been associated with climatic parameters in other countries (Xu et al., 2011), and in Mongolia historical incidence is loosely associated with annual rainfall perhaps because rainfall is conducive to an increased rodent population and, therefore, an increase in contact between rodents, their flea ectoparasites and humans.

- **Waterborne diseases**

Waterborne diseases such as dysentery, salmonellosis, typhoid and cholera account for one of the largest environmentally-mediated burdens of disease in the Western Pacific Region and cause significant morbidity in children under five years of age. In many instances, transmission of waterborne diseases is facilitated by water scarcity and the unavailability of clean water for drinking, washing and maintenance of good hygiene and is compounded by poverty, which is closely linked to these determinants. However, to compound the issue of water scarcity, increased rainfall following extensive droughts prevents proper water absorption into the ground and into smaller rivers and streams, which provide the main source of water for most populations. Inadequate absorption also results in surface runoff and the pollution of drinking-water sources. This water pollution has been linked to the increased incidence of diarrhoeal diseases and a considerable body of evidence links climate change impacts such as temperature and rainfall with altered incidence of these diseases (WHO, 2005; McMichael et al., 2006). In developing countries, for example, it has been inferred that warming of 1 °C will be associated with an increase in diarrhoea of 5% (WHO, 2004). In Bangladesh, the number of cholera cases increased with both high and low rainfall, albeit with different lag times (Hashizume et al., 2008). These events were also dependent on broader regional climatic phenomena: both negative and positive Indian Ocean Dipole (IOD) events are associated with increased disease incidence, with varying time lags (Hashizume et al., 2011).

Projections of future climate impact on health outcomes including diarrhoea are beset by uncertainties around climate forecasts, health impacts and the degree to which future socioeconomic changes will modify impacts. A recent paper used the results from a number of empirical studies and climate models to understand the range of uncertainty surrounding climate change impacts on diarrhoea, concluding that despite considerable uncertainty due primarily to a lack of empirical climate-health data, the impact of climate change will be substantial (Kolstad & Johansson, 2011).

In addition to diarrhoeal illnesses and typhoid fever, there are a number of other water-sensitive diseases endemic to the Western Pacific Region, the burden of which is likely to be affected by altered rainfall patterns and extreme weather events in the face of climate change. Such diseases include arsenicosis, leptospirosis, melioidosis, schistosomiasis and viral hepatitis (Meng et al., 2011).

2.4.4 Increased incidence of malnutrition

Malnutrition is one of the common consequences of climate change in vulnerable populations. Climate change can affect nutrition either in direct or indirect ways through its impact on food availability, stability of food supplies, access to food and food utilization. Although the increasing concentration of ambient CO_2 and rising temperatures can positively affect total crop yields, increased episodes of extreme weather events and changes in rainfall patterns, such as a prolonged monsoon or drought, often have a negative impact. Relocation from rising sea levels or natural disasters can limit access to food for affected people.

Less privileged population groups, defined by age, gender or socioeconomic status, are more vulnerable to malnutrition in food shortages induced by extreme weather events. Children are especially vulnerable to malnutrition because adequate nutritional supply is crucial to meet the need for the growth and development in the early stages of life.

In many countries in the Western Pacific Region, malnutrition is a major health issue in children, although rapid progress is being made. In the Western Pacific Region, some countries have a high prevalence of food insecurity, affecting 38.3% of people in the Lao People's Democratic Republic, 27.1% in Cambodia, 23.8% in the Philippines and 22.9% in Papua New Guinea (FAO, 2013). Stunting in children under five is common in these countries, ranging between 30% and 48%. Approximately 10% of children are affected by wasting, and between 20% and 30% of the children are underweight (FAO, 2013).

Extreme weather events are common precipitators of malnutrition. In the Lao People's Democratic Republic, areas affected by floods and typhoons showed a significant increase in acute malnutrition (Ministry of Health, Lao People's Democratic Republic, 2011). In a rural village of highland Papua New Guinea, annual average birth weight decreased in 1982 and

Figure 12. Effect of *dzud* on children's nutritional status in Mongolia

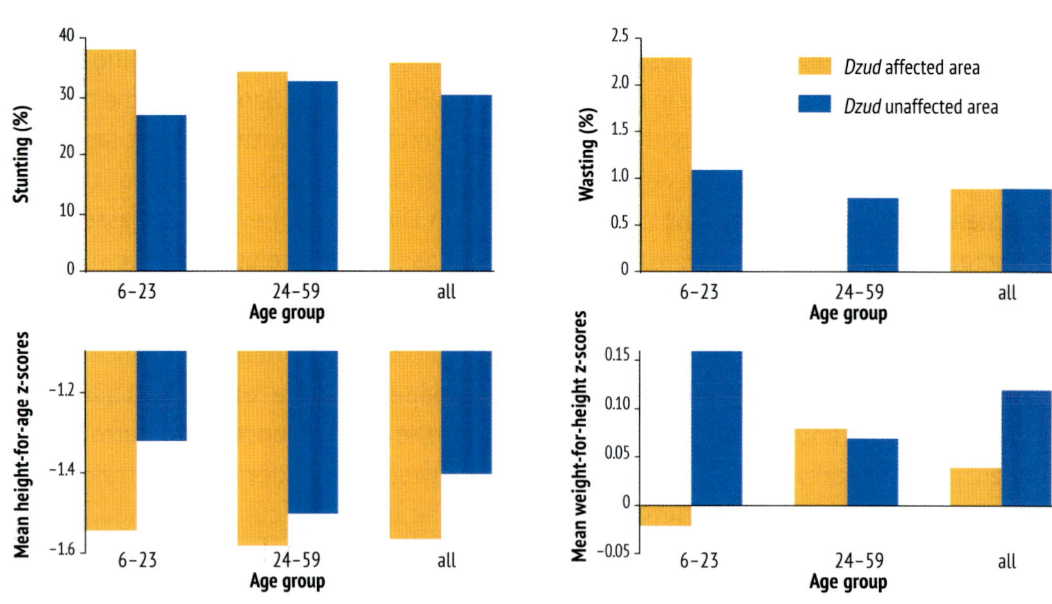

Source: drawn based on data from United States Centers for Disease Control and Prevention, 2001.

1983, which corresponded to a period of highest El Niño activity. In traditional highland communities of Papua New Guinea, children and women have the lowest priority for access to food, thus were the most immediately affected by the relative food shortage due to a long-lasting El Niño-pattern drought (Allen, 2002).

Severe cold winter weather also aggravates food shortages for children. After consecutive occurrence of the *dzud* (a severe cold and dry winter) in Mongolia during 1999–2001, massive loss of livestock was reported over a wide geographical range of the country. A survey on the nutritional status of children in affected areas demonstrated higher rates of stunting, loss of weight and lower haemoglobin among children in the affected area. This effect was evident only in children under two years (Fig. 12) (United States Centers for Disease Control and Prevention, 2001).

CHAPTER 3

Synthesis of climate change and health of selected countries in the Western Pacific Region

The aim of this report is to bring together information, experiences and best practices in the Western Pacific Region as a repository from which Member States and other stakeholders can draw on climate change and related health programmes. Seven countries included for discussion – Cambodia, the Lao People's Democratic Republic, Mongolia, Papua New Guinea, the Philippines, the Republic of Korea and Viet Nam – were selected to reflect the diversity of the Region.

In this chapter, seven topics will be covered: i) geography, population and health status; ii) GHG emissions; iii) future climate projections; iv) health risks related to climate change; v) vulnerability assessments; vi) governance and national activity on climate change and health; and vii) health adaptation activities.

For each country, a literature review was conducted to identify strategies, plans and frameworks for climate change and health. These include national plans for climate change and health, as well as more general climate change commitments that may include health components. Background and projections on climate were extracted from publicly-available sources according to IPCC scenarios (IPCC, 2000). Adaptation activities were described in literature and reports, and in some cases were already known to the authors or were gleaned from national colleagues and focal points. Authors have endeavoured to be comprehensive in summarizing activities but take responsibility for omissions.

3.1 Cambodia

> **"Cambodians are highly vulnerable to the health impacts of climate change. Most regions in Cambodia have limited adaptive capacity to respond positively to the impacts of climate change given high levels of poverty, low educational levels, technological and infrastructure limitations, and issues of governance.**
>
> Ministry of Health Cambodia and WHO, 2010

3.1.1 Geography, population and health status

Cambodia borders the Lao People's Democratic Republic, Thailand and Viet Nam and has a southern coast on the Gulf of Thailand. The country features central plains on which lie the Mekong River and the Tonle Sap Basin, with mountains and highland areas to the west, east and north. Approximately 80% of the 13.4 million-strong population lives in rural areas, highly dependent on agriculture dominated by paddy farming (Ministry of Environment Cambodia & United Nations Development Programme [UNDP] Cambodia, 2011). Urban migration has become a recent reality; since 1998, the proportion of the population living in the capital city of Phnom Penh has doubled, fuelling demographic, developmental and health-care challenges that are only now beginning to emerge (Ministry of Planning, Cambodia, 2012).

The health status of the Cambodian people has steadily improved over recent decades in line with impressive, if unequal, economic growth. Nonetheless, challenges remain. Infectious diseases constitute the main causes of mortality and morbidity, including acute respiratory infections, gastroenteric infections, and outbreak-prone waterborne and vector-borne diseases. The country is classified as high-burden for tuberculosis. NCDs and injuries will be the challenge of the future: surveys have indicated high levels of diabetes and hypertension in rural and urban areas, and the number of road accidents is rising rapidly. Health risks are exacerbated by environmental circumstances, especially the lack of safe drinking-water and poor sanitation and hygiene (WHO, 2011b).

3.1.2 Greenhouse gas emissions

A projection analysis of GHG emissions and removals by sector in 2000 indicated that Cambodia was already a net emitter of GHGs and these are likely to increase seven-fold by 2020. Among sectors, land use change and forestry (LUCF) is projected to be the main source of GHG emissions (63.0%), followed by agriculture (27.5%), with energy contributing only 9.0% of the national total (Ministry of Environment Cambodia, 2002).

3.1.3 Future climate projections

Cambodia has two distinct seasons: a dry season from mid-November to April and a rainy monsoon season from May to October, interrupted by a short "mini dry season" in late July and August. The annual average temperature is 28 °C, with average maxima and minima of 38 °C in April and 17 °C in January, respectively. The average annual rainfall is 1400 mm in the central lowland regions but may reach 5000 mm in coastal zones. The ring of mountain ranges affords the country protection from severe storms or cyclones, but when extreme weather does occur it predominantly strikes coastal regions from August to November. Floods commonly occur between May and October, which result from heavy rains that fall both locally and upstream in the Mekong Basin. Widespread drought occurred throughout the country in 1986 and 1987 and in 1997 and 1998. Both floods and droughts have caused considerable economic losses and associated social and environmental impacts (Ministry of Environment, Cambodia & UNDP, 2011).

Cambodia's temperature has been rising at a rate of approximately 0.18 °C per decade over the past 50 years and further increases are predicted until 2100.

The main climate change phenomena expected to occur in Cambodia include:

- increased temperatures, with corresponding increases in evaporation and transpiration;
- increased frequency and intensity of extreme events, such as floods and droughts;
- changes in seasonal distribution of rainfall, with drier and longer dry seasons, and shorter, more intense wet seasons;
- increased volume and intensity of wet-season rainfall, leading to increased floods and a marginal decrease in dry-season rainfall; and
- reduced flow of the Mekong and its tributaries in the dry season and increased flow in the wet season (Ministry of Environment, Cambodia & UNDP, 2011).

3.1.4 Health risks related to climate change

The Ministry of Health's *Climate Change Strategy for Public Health* (CCSPH) identifies three key areas of concern in the context of climate change and health in Cambodia: vector-borne diseases, waterborne and foodborne diseases, and the health impacts of extreme weather events (Ministry of Health, Cambodia, 2012). These priorities are based on the national climate change and health vulnerability and adaptation assessment, which was completed by Ministry of Health with support from WHO in 2010 (Ministry of Health, Cambodia & WHO, 2010).

- **Vector-borne diseases including malaria and dengue fever**

The incidence of vector-borne diseases such as dengue is extremely seasonal in Cambodia, and variations in weather patterns such as rainfall and temperature may impact the epidemiology of the disease (Fig. 13) (Ministry of Health, Cambodia, 2010). Variables include the control of breeding sites for mosquito vectors, the developmental time of vectors and the incubation period of the virus. In addition, human factors such as increased urbanization, international travel and migration also impact the epidemiology of the disease. Still, incidence is projected to increase. Impacts of climate change on malaria are possible but are likely to be obscured by successful and intensive malaria public health campaigns, including the distribution of insecticidal bednets and the timely use of effective antimalarial treatments. Indirect pathways, including food insecurity, population mobility and the availability of public health infrastructure, may exert additional impacts on vector-borne disease incidence, morbidity and mortality.

- **Waterborne and foodborne diseases**

In Cambodia, diarrhoeal disease is a very significant cause of morbidity and mortality, particularly in children. The Ministry of Health's Department of Planning and Health Information monthly reports show that diarrhoeal illnesses constituted the second-most common outpatient and inpatient diagnosis in 2013, with acute respiratory illnesses being the most common. Despite the limited diagnostic capacity in Cambodia, previous research findings suggest that the common etiologies of paediatric diarrhoeal diseases in Cambodia include *Escherichia coli* and rotavirus, with *Shigella* species implicated in cases of dysentery (bloody diarrhoea) (Meng et al., 2011). Cholera also occurs in relatively frequent epidemic cycles in Cambodia and has been shown to be strongly linked to changes in temperature, rainfall and other environmental conditions (Hashizume et al., 2008; Jutla et al., 2011).

However, in addition to diarrhoeal disease, there is a much longer list of diseases that are transmitted by water via contact, ingestion, inhalation, skin penetration and other means, or are otherwise affected by water (for example, contamination of potable water supplies, altered geographic range or habitat of vectors or vertebrates that transmit zoonoses, and interaction with soil saprophytes to bring them closer to the surface) and are thus susceptible to climate change. In Cambodia, examples of diseases that have been shown to exist include

Figure 13. Relationship between monthly rainfall, temperature and dengue incidence in selected Cambodian provinces

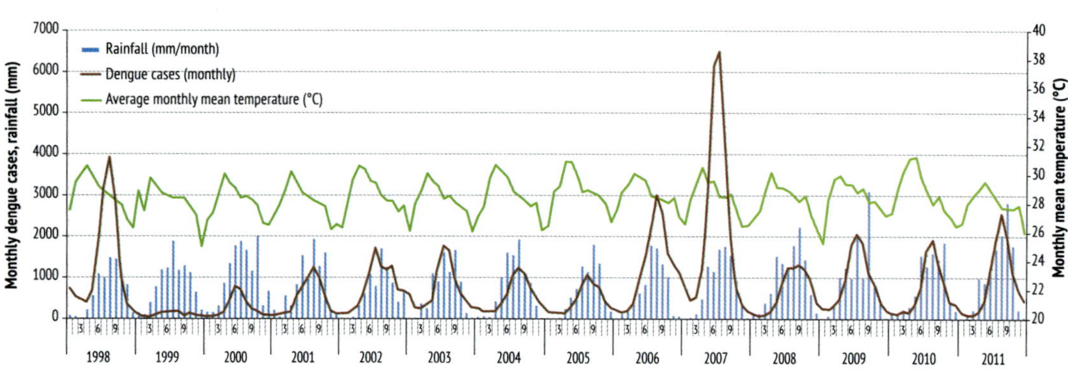

Source: Ministry of Health, Cambodia, 2010.

typhoid fever (Kasper et al., 2010; Wijedoru et al., 2012), leptospirosis (Berlioz-Arthaud et al., 2010; Ivanova, et al., 2012; Laras et al., 2002; Seng & Sok, 2007; Victoriano et al., 2009), melioidosis (Pagnarith et al., 2010; Rammaert et al., 2011; Vlieghe et al., 2011; Wuthiekanun et al., 2008), hepatitis E (Kasper et al., 2012), schistosomiasis (Muth, et al., 2010) and arsenicosis (Buschmann et al., 2007; Feldman et al., 2007). Again, there is an abundance of evidence linking most of these diseases to changes in climatic conditions, with significant international concern mounting over the prospect of increasing burdens of these diseases with climate change, particularly for leptospirosis (Desvars et al., 2011; Lau et al., 2012) and melioidosis (Inglis et al., 2009).

Certain subsectors of Cambodia's population may be considered to be more vulnerable than others with respect to climate change impacts on water and health. These groups include residents of flood- and drought-prone areas (Few et al., 2004) and certain occupations, for example, rice farmers and other agricultural workers. Both of these groups may be considered at increased risk of exposure to diseases transmitted via contact with pooled water, such as that which occurs in rice paddies or during flood conditions, notably diarrhoeal disease and leptospirosis (Cann et al., 2013).

Preliminary findings from Developing Research and Innovative Policies Specific to the Water-related Impacts of Climate Change on Health – a joint project by WHO and the Cambodia Ministry of Health – suggest that some provinces in Cambodia adjacent to the country's major river systems experience an increase in cases of diarrhoeal disease with increasing rainfall, while coastal provinces experience the opposite association (Ministry of Health, Cambodia & WHO, 2010).

People living in poverty may also be considered to be at higher risk (Nuorteva, Keskinen & Varis, 2010), largely due to their limited ability to access improved water and sanitation facilities. A "knowledge, attitudes and practices" survey carried out by the Ministry for Rural Development in 2010 found a strong correlation between households that had latrines and those that treated water appropriately and practised safe hygiene. This suggests that factors such as income and education are very likely strongly linked with health-protective behaviours, such as latrine use, water treatment and hand hygiene. Similarly, a 2003 World Bank report on the poverty–environment nexus in Cambodia showed a close statistical and spatial correlation between poor households and lack of access to safe water, with discussion of the implications for childhood diarrhoea and mortality (World Bank, 2013).

Changing rainfall and temperature patterns are expected to lead to challenges of providing adequate clean water, sanitation and drainage. While increased rainfall may reduce water scarcity in some regions and increase the availability of fresh water, the challenge of storing water safely between the wet and dry seasons remains: where facilities are inadequate, improper long-term water storage may be unsafe. In some regions, reductions in rainfall will lower river flows, reducing effluent dilution and leading to increased pathogen loading in freshwater supplies.

Changes to Himalayan glacial melts will also affect water availability in the future: increased melting will cause greater flows in the Mekong River and its tributaries that may lead to flooding events. However, in the longer term, annual river flows may dramatically decrease. These events have important health implications because 86% of Cambodians meet their water needs from the Mekong River Basin. Any alterations to the flooding cycle of the Tonle Sap system will also have impacts on human health for a number of reasons including depletion of important fish stocks, increased consumption of contaminated food facilitated by increasing ambient temperatures, and increased transmission of pathogens by insects and rodents.

- **Food security**

Climate change will affect all four dimensions of food security: food availability; access to food; stability of food supplies and food utilization, with consequent food security concerns. Increasing carbon dioxide levels, ambient temperatures and altered water availability in a changed climate will influence food production, availability, accessibility and quality through the direct effect on crop yields.

Some impacts may result in greater yields, such as increases in average temperature by 1–3 °C. However, temperature increases beyond 3 °C are likely to result in decreases in production. Changes in precipitation patterns will affect rice yields in Cambodia, as the crops are predominantly rain-fed. An additional concern is the impact of global food production changes as a result of climate change. Prices and trade are likely to be affected with particular impacts on developing countries.

- **Health consequences of extreme weather events**

Over recent years, an increasing frequency of severe flooding events has been observed in Cambodia and flood periodicity has also increased, affecting large populations almost annually over the past decade. Additionally, the pattern of flooding has changed in several provinces including Kandal, Kampong Cham, Kampong Chhnang, Kampong Thom and Takeo. The sea level has been rising at 2–3 mm per year over the past two decades, and is projected to accelerate to a rate of about 5 mm per year over this century. This will increase the number of people in Cambodia's coastal areas at risk from flooding and saltwater intrusion, which may be exacerbated by declining dry-season precipitation. Associated effects may be felt by industry and agriculture in coastal areas. The health consequences of extreme weather events are well documented and include increased morbidity and mortality from heatwaves, floods and droughts; increased burdens on health services; food shortages as a result of crop destruction, leading to malnutrition; effects on water supplies, sanitation and drainage; and mental health impacts.

3.1.5 Vulnerability assessment

The low-lying nature of Cambodia's topography, coastal exposure, proximity of densely populated regions to flood- and drought-prone areas, and the demographic profile all increase the sensitivity of the population to the health impacts of climate change. Most regions have limited adaptive capacity to respond positively to the impacts of climate change given high levels of poverty, low educational levels, technological and infrastructure limitations, and issues of governance. Furthermore, coastal populations in the country's south are at risk from sea level rise.

Urbanization is under way in Cambodia, and this phenomenon is likely to increase under future climate change conditions, as employment and livelihoods in rural areas become more challenging. Infrastructure, including sewerage, roads and housing, and easy access to educational and health services are often lacking at the burgeoning edges of cities and large towns. Undernutrition among urban Cambodian children is now recognized as a major issue, and urban populations are strongly affected by rural crop failures or other agricultural shortfalls. The most vulnerable populations in Cambodia for food insecurity are children, older people, pregnant women and those with chronic illness (Ministry of Health, Cambodia & WHO, 2010).

The relative scarcity of health resources and health professionals in Cambodia also increases the sensitivity of the population to the more severe health impacts of climate change, and the adaptive capability of the health sector to meet the challenges of climate change in Cambodia is limited.

3.1.6 Governance and national activity on climate change and health

In June 2003, the Cambodian Government established the Climate Change Department (CCD), which is solely dedicated to climate change issues and is embedded within the Ministry of Environment. CCD is responsible for carrying out all technical activities related to the implementation of the UNFCCC and related international conventions, in addition to facilitating and coordinating donor-funded and private sector activities. CCD also supports and organizes interministerial technical working groups specializing in various sectors (Fig. 14) (Joint Climate Change Initiative, 2010).

The National Climate Change Committee (NCCC) is an interministerial mechanism with the mandate to prepare, coordinate and monitor the implementation of policies, strategies, legal instruments, plans and programmes of the Government to address climate change issues within the country. The NCCC is chaired by the Prime Minister and is composed of the secretaries and under-secretaries of state from 19 ministries and government agencies (Joint Climate Change Initiative, 2010).

Figure 14. Cambodia's national climate change management structure

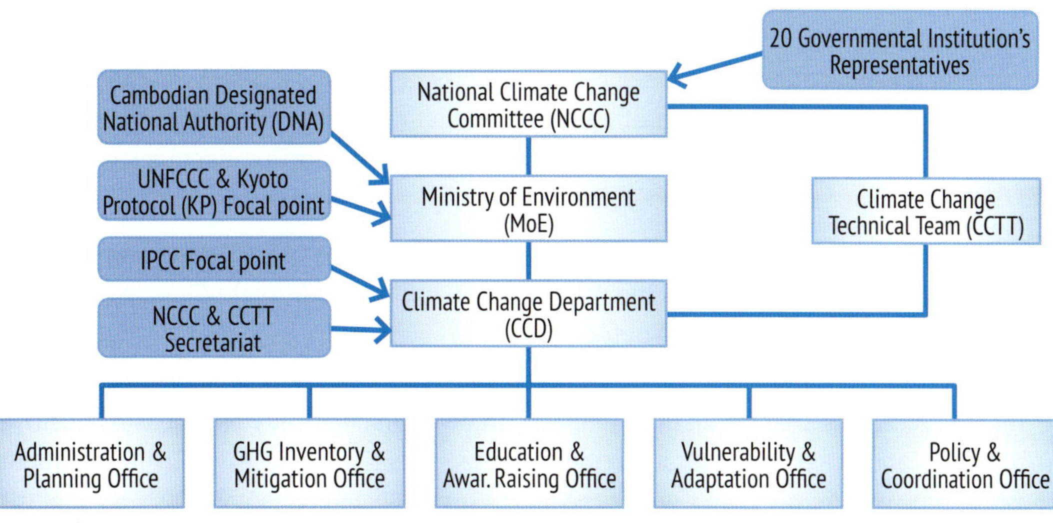

Source: Joint Climate Change Initiative, 2010.

The Ministry of Health is guided by its *Second Health Sector Strategic Plan 2008–2015* (HSP2) and plays a cross-cutting role in climate change adaptation, including activities related to gender and the health of minorities. The Department of Preventive Medicine is the focal point for climate change and health, and represents the health sector on the NCCC where the health impacts of climate change initiatives can be considered and health planning mainstreamed into climate change adaptation activities (Ministry of Health, Cambodia, 2008). A new *Climate Change Strategic Plan for Public Health*, developed by the Ministry of Health with a broad range of inputs, is the health sector's contribution to the *Cambodia Climate Change Strategic Plan,* and will guide national adaptation plans in the future (Ministry of Health, Cambodia, 2012). The plan aims to provide a policy response to climate change and health risks and focuses on three priority areas:

- improving health-care infrastructure and the capacity of health personnel to cope with vector-borne and waterborne diseases in the context of climate change;
- enhancing emergency preparedness and response to cope with extreme weather and climate change-related disasters; and
- improving the knowledge and research capacity on health impacts and vulnerability to climate change as an information base for mainstreaming climate change in the health strategic planning activities of the Ministry of Health and other sector planning.

3.1.7 Health adaptation activities

Adaptation measures in Cambodia have been guided by the 2010 Climate Change and Health Vulnerability and Adaptation Assessment. A project to strengthen the control of vector-borne diseases was implemented by the Ministry of Health with WHO technical support in an effort to lessen the impacts of climate change. The project, implemented between 2011 and 2012 and funded by the Korean International Cooperation Agency (KOICA), included strengthening disease surveillance systems, supporting mosquito vector surveys to determine the seasonal and geographical distributions of vector species, and strengthening clinical and outbreak response capacity in identified vulnerable areas. Community-based activities raised awareness among vulnerable population groups, and research was conducted into the causes of disease, including a review of historical records. As a result, health workers and the population in vulnerable communities are aware of climate change and vector-borne disease risks and response strategies which have been incorporated into national policies and frameworks (WHO, 2012b). The project was conducted in close cooperation with the environmental and meteorological sectors, among others, and was a successful intersectoral example of climate change and health adaptation. Elements of the project were extended as part of a Cambodian Climate Change Alliance Initiative, supporting dengue surveillance and outbreak response in 2013 and 2014.

In addition to the work on climate change and vector-borne diseases, the Developing Research and Innovative Policies Specific to the Water-related Impacts of Climate Change on Health project has achieved the following outcomes:

- increasing the awareness of health professionals and other stakeholders in Cambodia regarding climate change, its likely impacts on health (specifically on diseases sensitive to water), and the presence of a number of water-sensitive diseases in Cambodia about which little is currently known within the health sector (including leptospirosis, melioidosis and schistosomiasis);

- increasing the knowledge of health professionals and other stakeholders in Cambodia regarding the application of environmental epidemiological techniques to public health practice;
- increasing the capacity of the health professional community in Cambodia (Ministry of Health and WHO) to apply Geographical Information System (GIS) technology and spatial analysis for public health;
- increasing the understanding of the relationship between historical climate variables (such as temperature, rainfall and river height) and the incidence of diarrhoeal disease in Cambodia;
- increasing the understanding of other factors that contribute to the burden of diarrhoeal disease in Cambodia, such as geographic location, socioeconomic status and access to improved water and sanitation facilities;
- demonstrating the "climate sensitivity" of waterborne diseases in Cambodia, particularly diarrhoeal disease; and
- compiling health promotion materials aimed at increasing community awareness regarding the risks posed by climate change to water and health, and educating the public about strategies to reduce those risks.

3.2 Lao People's Democratic Republic

> **"Almost all the Lao population living in rural areas is heavily dependent on forests for their subsistence, income generation, energy, and agriculture and rural development."**
>
> Food and Agriculture Organization of the United Nations webpage

3.2.1 Geography, population and health status

The Lao People's Democratic Republic is a landlocked country stretching more than 1700 km from north to south and up to 400 km from east to west. It shares borders with Cambodia, China, Myanmar, Thailand and Viet Nam. Mountains cover about two thirds of the land area, while several rivers including a 1856 km stretch of the Mekong River (Government of the Lao People's Democratic Republic, 2000) traverse it. The country has a predominantly rural population of 6.1 million people, sparsely populated, but with large variations among its 17 provinces (WHO, 2011b).

NCDs accounted for 48% of all deaths in the Lao People's Democratic Republic in 2008, while communicable diseases including malaria and tuberculosis, as well as maternal and perinatal and nutritional conditions, accounted for 41% of deaths. Chronic malnutrition remains an alarming challenge with 26.6% of children under 5 underweight and 44% stunted. In rural areas, every second child is malnourished. Additional and growing problems include road accidents, which are increasing in line with higher volumes of traffic, and mental health issues, particularly drug abuse (WHO & the Ministry of Health, Lao People's Democratic Republic, 2012; WHO, 2011a).

3.2.2 Greenhouse gas emissions

The Lao People's Democratic Republic remains a net CO_2 sink, predominantly as a consequence of forest growth (Lao People's Democratic Republic, 2009). The agricultural sector does give rise to methane production: total emissions in 2000 were 312 gigagrams (Gg, equal to 1000 tonnes), of which agriculture accounted for 81%. CO_2 emissions are primarily from onsite burning of wood in forests and from traditional biomass. The NO_x emissions in the economy are negligible (Government of the Lao People's Democratic Republic, 2000, 2009). As the capital city of Vientiane continues to develop, it remains to be seen whether increasing vehicular traffic will cause substantial GHG emissions or air pollution at levels sufficient to impact human health.

3.2.3 Future climate projections

The Lao People's Democratic Republic has a tropical climate, which is influenced by the south-east monsoon, causing significant rainfall and high humidity. There are two distinct seasons: rainy, or monsoon, and dry. Annual average rainfall and temperatures vary from approximately 1300–3000 mm and from 20–27 °C depending on geographical area (Ministry of Health, Lao People's Democratic Republic, 2011). The Lao People's Democratic Republic is divided into three different climatic zones on the basis of altitude that differ in terms of temperature, rainfall, and extreme event frequency and magnitude: northern mountainous areas above 1000 m that are dry and cool; central mountainous areas with a warm, tropical monsoonal climate; and highly populated, tropical lowland plains and floodplains along the Mekong River and its main tributaries that have an annual average rainfall of 1500–2000 mm (Lao People's Democratic Republic, 2009).

Projected climate change impacts in the Lao People's Democratic Republic are likely to vary significantly among regions and available data at the subnational level are limited. Models generally predict the climate in the Lao People's Democratic Republic will warm 1.4–4.3 °C by 2100, and that dry seasons will become longer. However, there is also a likely increase in rainfall across eastern and southern areas of the country, by up to 10–30% including an increase in the intensity and frequency of extreme events such as flooding (World Bank Group, 2011a). Additionally, Mekong-wide modelling studies have indicated that while average temperatures are unlikely to change significantly under moderate emissions scenarios, the number of hot days is likely to increase and the number of cool days decrease, both significantly. Wet days are also likely to increase (Lao People's Democratic Republic, 2009).

Figure 15. Monthly rainfall in each of three Lao regions and nationally, 1998–2010

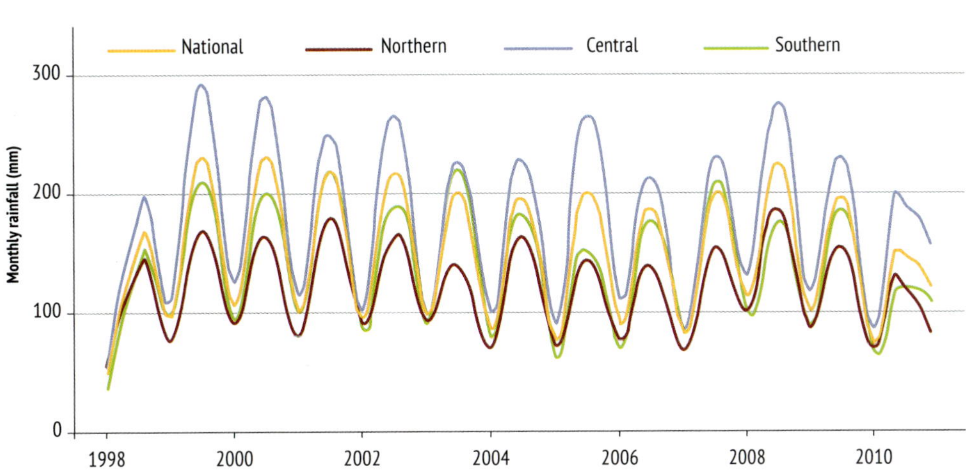

Source: Ministry of Health, Lao People's Democratic Republic, 2011.

3.2.4 Health risks related to climate change

A climate change and health vulnerability assessment was carried out in the Lao People's Democratic Republic by the Ministry of Health with technical assistance from WHO in 2011, using retrospective data in addition to an assessment of current burdens and capacities (Ministry of Health, the Lao People's Democratic Republic, 2011).

- **Communicable diseases**

Dengue is seasonal in the Lao People's Democratic Republic, with most cases occurring in the rainy season (Imai et al., 2013). While the determinants of outbreaks are complex and include both intrinsic and extrinsic factors, cases might be expected to increase following increases in humidity and temperature. Human activities, including altering patterns of migration, urbanization and water storage, have been observed to impact dengue incidence elsewhere and may be the subject of future study in the Lao People's Democratic Republic. The incidence of waterborne diseases, particularly diarrhoea and dysentery, has been increasing in recent years. The association of diarrhoea with drought is a concern: climate projections are that dry spells will increase. Incidence of waterborne diseases (such as dysentery, typhoid fever and hepatitis A) is also associated with climate and strengthened surveillance may provide evidence of association between climate change and their incidence.

- **Water and sanitation**

The Lao People's Democratic Republic has abundant freshwater supplies, but climate change may place these in jeopardy. The 2009 NAPAs prioritize drinking-water and sanitation, particularly in flood- and drought-prone areas, as important areas for adaptation. Specific actions include enhancing the capacity of engineers, raising awareness particularly in vulnerable areas, and the prevention and treatment of waterborne diseases. Strengthened surveillance, diagnosis and laboratory capacity, in addition to the communicable and epidemic-prone diseases described above, are also adaptation priorities (Lao People's Democratic Republic, 2009).

- **Nutrition and extreme weather events**

Food shortages and malnutrition following typhoons, flooding and other extreme weather events are projected to increase. For example, Typhoon Ketsana in September 2009 affected over 180 000 people in five southern provinces of the country, wreaking an economic loss of US$ 1.25 million to public health facilities and equipment alone. Flooding of low-lying areas is a particular risk: 93% of those impacted by floods had their food source affected. Substantial resources were required to provide emergency health care to affected populations. Injury and death following extreme weather events is also a concern given projections of increased instability and increased frequency of typhoons (Government of the Lao People's Democratic Republic, 2009).

3.2.5 Vulnerability assessment

Health vulnerability to climate change in the Lao People's Democratic Republic is predominantly a product of exposure to drought, flooding, deforestation and the loss of biodiversity. As with other countries, those most affected are individuals most dependent on an agrarian lifestyle and least able to adapt, including farmers, unskilled labourers, fishermen, and those reliant on the forests for hunting and gathering. All rural areas are vulnerable, with more specific vulnerabilities in areas liable to flooding and in particular, drought.

The population of the Lao People's Democratic Republic, possibly more so than any other country in the Western Pacific Region, is closely integrated with the natural environment. Highland populations gather non-timber products from the forest and use traditional cultivation techniques, both of which are activities threatened by impacts of climate change in combination with increased environmental pressure tied to other factors. These impacts have contributed to deterioration in nutritional and health status, and affected populations should be considered particularly vulnerable (Lao People's Democratic Republic, 2009; Ministry of Health, Lao People's Democratic Republic, 2011; WHO, 2011b).

3.2.6 Governance and national activity on climate change and health

The Lao People's Democratic Republic ratified the UNFCCC in 1995 and the Kyoto Protocol in 2003. The Climate Change Office, Ministry of Natural Resources and Environment, was established in 2008 to serve as the secretariat of the National Steering Committee on Climate Change, since replaced by the National Environment Committee. It acts as the national focal point on climate change actions and initiatives, and coordinates a number of the national government's activities related to the UNFCCC. Various sectoral interests are represented by technical working groups (Sengchandala, 2010). In 2010 the Government published the *Strategy on Climate Change of the Lao People's Democratic Republic*, describing risks and adaptation options including in the area of public health. Its focus was water and sanitation, communicable diseases, awareness-raising, and streamlining and strengthening existing programmes and structures (Watershed Resource and Environment Administration of the Government of the Lao People's Democratic Republic, 2010). Taking a regional approach, the Climate Change and Adaptation Initiative of the Mekong River Commission (CCAI) is a collaborative regional initiative of Cambodia, the Lao People's Democratic Republic, Thailand and Viet Nam, aiming to support adaptation to the impacts and new challenges posed by climate change through improved planning, implementation and learning (Mekong River Commission, 2011).

Within the Ministry of Health, climate issues are addressed by the Environmental Health Division, which also oversees water supply quality, sanitation and hygiene, and the water and sanitation sector assessment. The division's involvement in adaptation is built around the *Climate Change and Health Adaptation Strategy in the Lao People's Democratic Republic* prepared in conjunction with WHO, and an accompanying five-year action plan (Ministry of Health, Lao People's Democratic Republic, 2011).

3.2.7 Health adaptation activities

The Lao People's Democratic Republic health adaptation strategy includes a climate change and health vulnerability assessment study carried out in 2011, the strategy itself and a five-year action plan (2012–2016). The objectives are to assess climate change impacts, improve disease monitoring systems and the control of infectious diseases, prepare and respond to food emergencies and to extreme weather events, strengthen health education and communication, and empower people to take actions to reduce individual and community vulnerability to climate change. Funding is required to implement the action plans.

3.3 Mongolia

3.3.1 Geography, population and health status

With a land area of 1.6 million km^2 and a population of 2.78 million people, Mongolia is one of the most sparsely populated countries on Earth. It has a unique geography of steppes, deserts, mountain ranges, and dry lake-dotted basins. Administratively, Mongolia is made up of 21 *aimags* (provinces) and the capital city of Ulaanbaatar (WHO, 2010d). One third of the population lives in Ulaanbaatar and 37.5% are rural, many of them continuing to practise a nomadic lifestyle of livestock herding (United Nations Development Programme, 2011; WHO & Ministry of Health, Mongolia, 2012).

Mongolia is in a state of epidemiological transition whereby cardiovascular diseases, cancer, injuries and poisonings have increased, in contrast to declining mortality from communicable and respiratory diseases. Risk factors contributing to NCDs, including smoking, alcohol consumption, overweight and obesity, are prevalent. As of 2010, the leading causes of morbidity per 10 000 population were diseases of the respiratory system (1157), digestive system (881), genito-urinary tract (737), and circulatory system (708), and injuries and poisonings (470) (WHO & Ministry of Health, Mongolia, 2012; WHO, 2010d).

3.3.2 Greenhouse gas emissions

Mongolia entered the 1990s with a highly carbon-intensive economy and it remains among the world's top 10 coal-producing countries. Reduction of the country's carbon footprint remains a priority (United Nations Development Programme, 2011). A GHG inventory conducted in 2006 found that the energy sector was the major source of GHG emissions, generating 65.4% of the total, while the agriculture sector and land use change and the forestry sector contributed 41.4% and 13.3%, respectively. Other relatively minor sources include emissions from industrial processes (5.6%) and the waste sector (0.9%). CO_2 is the most significant GHG, at approximately 50% of emissions, followed by methane, over 90% of which results from livestock herding. There was a 6% increase in GHG emissions from 2005 to 2006, attributable to transport, manufacturing and construction. Emissions related to mining also increased.

From 1990 to 2006, there was an average annual reduction of 2.3% in GHG emissions in the country, but per capita emissions remain high compared to other developing countries because of the cold climate, the long heating season and the low efficiency of fuel conversion. From 2006 to 2030 average growth in GHG emissions is projected at 9.33% per year (Ministry of Nature Environment and Tourism, Mongolia, 2010).

3.3.3 Future climate projections

Mongolia has a harsh continental climate with four distinctive seasons, high annual and diurnal temperature fluctuations, and low rainfall. Because of the high altitude, it is generally colder than other countries of the same latitude. The minimum temperature is usually between −31 °C and −53 °C in January and the maximum between 28.5 °C and 42.2 °C in July. Average annual precipitation is low and differs from province to province, ranging from 38.4 mm in the south (Gobi region) to over 380 mm in some northern areas. Most rainfall is concentrated in June, July and August. Droughts in the spring and summer occur about once every five years in the Gobi region, and once every 10 years over most other parts of the country. Extreme weather events include the phenomenon known as *dzud*, which refers to extreme winter weather conditions combining drought, very heavy snowfall, and extreme cold and ice cover during which livestock are unable to find fodder, when large numbers of animals die due to starvation and the cold, with associated economic and human health effects (Ministry of Nature and the Environment, Mongolia, 2001).

A surface water inventory conducted in 2007 revealed that 852 rivers and streams out of 5128 had dried up, and severe desertification is extending over much of the country (Fig. 16; UNDP, 2010). The country is prone to other natural hazards including floods and forest fires, and human and epidemic animal diseases.

The effects of climate change in Mongolia, located deep inland and at high latitudes, are expected to be especially prominent. Projections developed in the IPCC Fourth Assessment Report using global climate models indicate an increase in mean temperature of 3–4.6 °C by 2099. Likewise, an increase in precipitation is predicted by all models of up to 16%.

Figure 16. Desertification in Mongolia

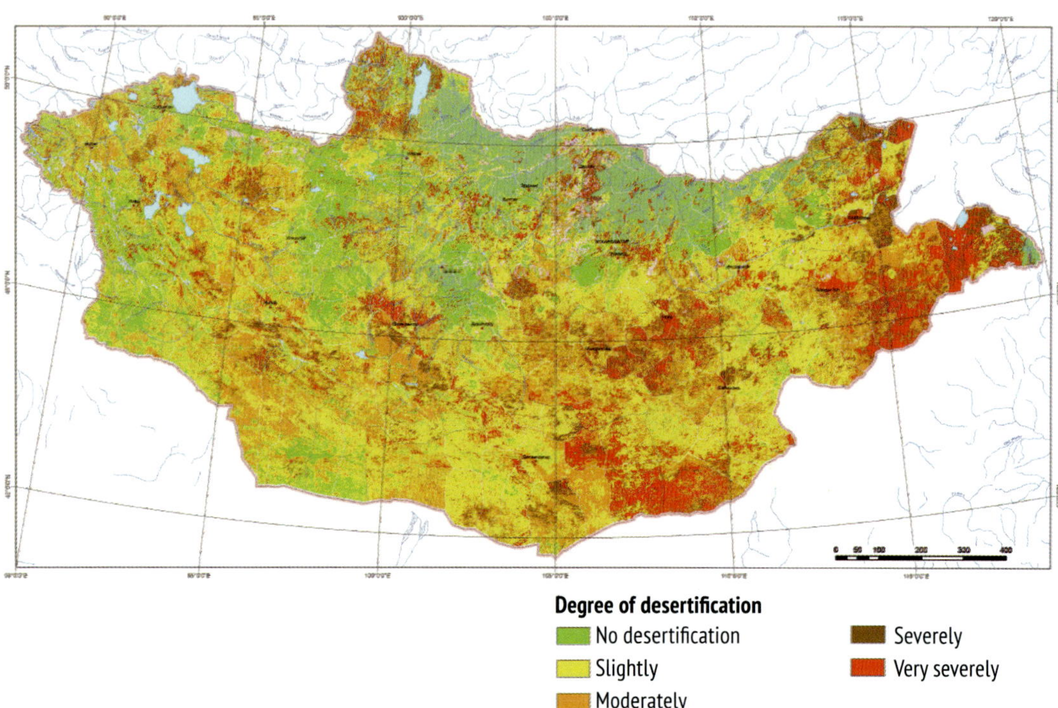

Source: Mandakh N., 2012.

The cumulative effects are likely to be milder winters with more snowfall, while summers will become hotter and drier, despite the increase in precipitation, due to higher surface evaporation that will obscure the impacts of rainfall. The severity and frequency of droughts are therefore expected to continue to increase through the 21st century. It is expected that the magnitude of warming will be higher in the summer (by 1.1–1.4 °C by 2011–2030, 2.7–3.6 °C by 2046–2065 and 3.7–6.3 °C by 2080–2099) than in the winter (increases of 0.2–0.7 °C, 1.6–2.5 °C and 3.0–3.8 °C, respectively) (Cruz et al., 2007).

3.3.4 Health risks related to climate change

An analysis of climate change and health was conducted in Mongolia in 2009 to understand the most prominent risks, identify vulnerable population groups and facilitate planning of appropriate adaptation responses with support from the WHO Regional Office for the Western Pacific (WHO, 2009b). The climate change and health risk factors are varied and include a number of indirect effects. For example, cardiovascular respiratory diseases cause significant morbidity and are seasonal; climate change may impact their incidence. Other risk factors for climate change and health include water quality and waterborne diseases, communicable diseases and the health consequences of extreme weather events.

- **Air quality and health**

Air quality in big cities, especially Ulaanbaatar, is worsening rapidly, raising a serious health threat. It is most prominent during the winter. In addition to coal combustion from the energy sector and a heavy traffic burden, the burning of fossil fuels from various sources, including traditional housing of incoming nomad populations, is responsible for the poor air quality during the winter. The influx of the nomad population is partly related to the loss of livestock from desertification and extreme weather events in the steppe. However, air quality monitoring was historically conducted on only two air pollutants, sulfur dioxide and nitrogen dioxide, and monitoring of particulate matter 10 (PM_{10}) began only recently in Ulaanbaatar. Mortality due to cardiovascular diseases has tended upward since the 1990s. There is a seasonal variation in respiratory and cardiovascular diseases, with high morbidity in winter. Correlation analyses show that angina pectoris is correlated with air pollution and respiratory diseases with air pollution and weather parameters, respectively.

- **Water quality**

Mongolia experiences considerable water stress as a result of insufficient and unreliable rainfall, changing rainfall patterns and flooding. The impacts of climate change are likely to add to this stress. Scarcity may be first noticed where populations depend on open water sources such as streams and springs, which are disappearing. Over recent years, an increasing trend of groundwater mineralization and saltwater intrusion has also been observed which may be exacerbated by climate change. Arsenic and fluoride contamination of groundwater sources in the Gobi region and microbial contamination in the urban and rural areas are major water quality issues. Major health impacts are likely to include increased incidence of waterborne diseases. These are related to both climatic determinants and water quality indices, all liable to change in future.

- **Communicable diseases**

Communicable disease incidence is associated with temperature, rainfall, humidity and other climatic patterns. Generally speaking, climate change projections for Mongolia are consistent with increased survival, propagation and outbreaks of human pathogens causing a number of human diseases. However, indirect impacts of climate change are more likely to affect disease incidence. Migration and unplanned urbanization as a consequence of pressure on traditional agrarian lifestyles, are predicted to facilitate disease transmission. Poverty and interruption to traditional lifestyles may also be associated with increased disease incidence. In Mongolia, vector-borne diseases include tick-borne diseases such as tick-borne encephalitis and Lyme disease. Increased tick distribution and disease incidence have been associated with a warming climate elsewhere, and these remain concerns in Mongolia (Jaenson & Lindgren, 2011).

- **Extreme weather events**

Common extreme weather events in Mongolia include strong windstorms, thunderstorms, heatwaves, droughts and flash floods, in addition to extreme snowfall and a combination of long, cold and dry winter periods, known as *dzud*. Projections indicate the frequency and intensity of these events will increase with a number of direct and indirect impacts on human health. Direct impacts include heatwave- and flood-related mortality, but indirect impacts are more likely and more serious due to effects such as the loss of livestock, desertification and other drivers of agricultural failure, as well as increased rates of diarrhoeal diseases following flooding.

3.3.5 Vulnerability assessment

Vulnerable groups in the country are those populations with least adaptive capacity and include the poorest populations most dependent on the environment for their daily needs. In Mongolia, nomadic herders make up a considerable proportion of the population and are particularly vulnerable. Children, the elderly and those with existing medical conditions are also highly vulnerable. Very specific population groups are vulnerable to specific threats. For example, those residing in forested areas are vulnerable to increased incidence of tick-borne diseases. More than half of the population is living in the capital area, and they are exposed to a very high level of air pollution during the winter, which is related to the extensive use of coal for heating and energy, as well as the influx of nomad populations from the countryside. Those in the arid south of the country, such as the Gobi region and eastern provinces, will be particularly impacted by future drought; existing water quality is poor and any deterioration may drastically impact their ability to survive in this environment.

3.3.6 Governance and national activity on climate change and health

The Government of Mongolia has established an interdisciplinary and intersectoral National Climate Committee (NCC), led by the Minister of Environment and Green Development and attended by high-level officials, to coordinate and guide national activities and measures intended to adapt to climate change and mitigate GHG emissions. The NCC approves the country's climate policies and programmes, evaluates projects, and provides guidance for these activities. Additionally, the Climate Change Office (CCO), established within the Ministry of Environment and Green Development, is tasked with carrying out day-to-day activities related to commitments under the Kyoto Protocol, implementing the guidance of the NCC and integrating climate change activities among sectors (Fig. 17) (Gomboluudev, 2007; UNEP & UNDP, 2009).

Since July 2012, climate change and health has been the responsibility of the Division of Public Health, guided by the *Strategy for Reduction and Adaptation to Climate Change and its Effects on Human Health (2011–2015)*, which was endorsed by the Minister of Health in 2011. Additionally, a thematic working group (TWG) on climate change and health was established by joint order of the Ministers of Health and Environment and Green Development in January 2011, focusing on coordination and implementation of activities in the country.

3.3.7 Health adaptation activities

In line with national climate change and health plans, the Ministry of Health with WHO support has initiated a number of health adaptation activities in Mongolia. In provinces at highest risk of vector-borne disease incidence, a KOICA-funded project was initiated in 2011 with a focus on strengthened surveillance and response for these climate-sensitive diseases. It was one of the first such projects conducted in the Western Pacific Region and brought together an intersectoral technical working group to ascertain baselines, build resilience within health

Figure 17. Organizational structure of climate change activities in Mongolia

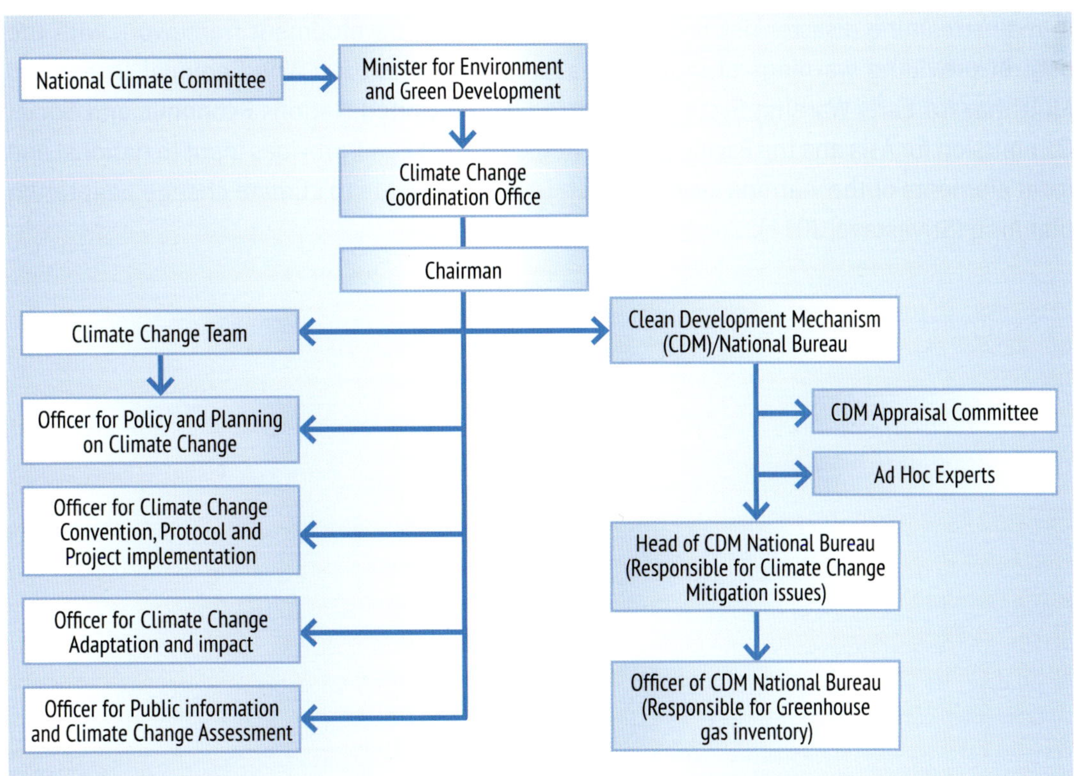

Source: Ministry of Environment and Green Development, Mongolia.

systems and strengthen intersectoral capacities. The cross-sectorality of activities was one of the project's strengths, reflecting the nature of climate change adaptation and the fact that many of these diseases are zoonotic. The environmental, veterinary and meteorological sectors were involved throughout the project. The initiative received a high level of political support and as a result, local and migrant populations are better informed about risks and protective behaviour, and they are able to protect themselves from being bitten by disease-transmitting insects. Health staff members are able to provide services directly to nomadic communities and research has been conducted to improve understanding of the complex relationships between climate, host animal behaviour, ectoparasite distribution and infectivity rates. The project contributed to attainment of the MDGs related to health (MDGs 4, 5 and 6) as well as those related to environmental sustainability and partnerships (MDGs 7 and 8) (WHO, 2012b).

An additional project integrated with water safety plans has been launched to adapt to the impacts of climate change on water scarcity. Supported by WHO, vulnerability assessments were conducted both on communicable diseases and NCDs in Dornogobi province, Gobi region, where water scarcity is most acute. As noted above, climate change is anticipated to negatively impact water resources and increase the number of years with scarce water. Populations in this area frequently utilize water containing excessive chemical elements including ammonium, arsenic, fluoride, iron and nitrite. Arsenic treatment facilities have been installed in the most vulnerable communities where arsenic levels exceeded WHO recommendations, and training on the proper dose of chlorine and monitoring of residual chlorine was conducted for water engineers.

Reflecting the high profile of the issue in Mongolia, WHO supports an annual national multisectoral symposium on climate change and health. The fourth symposium was held in October 2012 focusing on activities at the local health centre/hospital level to reduce environmental pollution and improve adaptation. Mongolia has also made excellent progress in mainstreaming disaster risk reduction into its overall development framework with the goal of providing warnings of multiple hazards, including *dzud*. The Regional Integrated Multi-hazard Early Warning System, supported by the United Nations Economic and Social Commission for Asia and the Pacific (ESCAP) Regional Trust Fund, provides input to national and local elements of the warning system which is of direct value to climate change adaptation planning (Srivastava, 2011).

3.4 Papua New Guinea

3.4.1 Geography, population and health status

Papua New Guinea is a tropical country occupying the eastern half of New Guinea island and nearby archipelagos. It is the biggest Pacific island country, with a total land area of 462 840 km^2. Its southern border faces Australia across the Torres Strait, the western border is Indonesia, and the eastern and northern marine territory faces Solomon Islands and the Federated States of Micronesia, respectively. A mountain range towering to 4500 m forms the spine of New Guinea island, clearly dividing south and north. There is no road connection between the capital, Port Moresby and other main cities in Papua New Guinea, only airline and marine connections. Papua New Guinea has an estimated population of 6.7 million people, 38.2% of whom are under the age of 15. Some 87.5% of the country's population lives in rural areas. Many rural areas have retained their traditional ways of life. The country is exceptionally diverse in its geography, ethnicity and language. Approximately 800 languages are spoken by various population groups and tribes. Administratively, the country has 22 provinces and 89 districts. Only half of women and two thirds of men over 15 years of age have attended school (WHO, 2011b).

Life expectancy is 62 years of age for men and 65 years for women, and Papua New Guineans have the lowest health status in the Pacific region. Maternal and child mortality are worsening, indicating a decrease in access to quality health services, and communicable diseases remain the leading causes of morbidity. Approximately 50% of deaths are caused by pneumonia, malaria, tuberculosis, diarrhoeal diseases, meningitis and HIV/AIDS. While detailed HIV prevalence data are lacking, AIDS is now the leading cause of death in adult inpatients at the Port Moresby General Hospital. NCDs are also strongly represented, including tobacco- and alcohol-related illnesses, diabetes, hypertension, and a number of cancers (WHO, 2011b).

3.4.2 Greenhouse gas emissions

Over 95% of Papua New Guinea's emissions derive from land use, and land-use change and forestry (LULUCF), including the effects of forest fires. The remaining emissions are from mining, transport and industry associated with the production of energy, oil and gas. The level of emissions is estimated at 115–131 Gg CO_2e (thousand tonnes of carbon dioxide equivalent) for 2010, which accounts for 0.01% of global GHG emissions. In February 2010, the Government of Papua New Guinea made a conditional commitment to the UNFCCC under the Copenhagen Accord, pledging that GHG emissions would be reduced by 30% from current levels, or 50% from the business-as-usual (BAU) forecast, by 2030 (Office of Climate Change and Development, Government of Papua New Guinea, 2010).

3.4.3 Future climate projections

Papua New Guinea has a monsoonal climate characterized by high temperatures and humidity throughout the year. However, in the highlands, the weather is mild with less seasonal variation in rainfall. The southern coastal area sees much less frequent precipitation and the climate is savannah. Although the temperature variation is quite small, there are distinct wet and dry seasons. Papua New Guinea has some of the wettest climates in the world: annual rainfall in many areas of the country exceeds 2500 mm, particularly in the highlands. Mean temperatures in the capital of Port Moresby tend to remain within the range 26–28 °C. Mean temperatures have increased across the Pacific by approximately 1 °C since 1970, and sea surface temperatures have increased by 0.6 °C to 1 °C since 1910. The El Niño Southern Oscillation plays a considerable role, and El Niño years are usually drier than normal. The sea level has risen by approximately 7 mm/year since 1993.

Climate projections for Papua New Guinea predict that temperatures will continue to increase, by up to 3.4 °C by 2090, according to a high-emission scenario. The number of very hot days is also predicted to increase. While future rainfall patterns are uncertain, it is likely that the number of extremely wet days will increase. An increased frequency of El Niño events may increase the level of drought, and increase the intensity of cyclones affecting coastal regions. In combination with predicted sea level rises of 4–15 cm by 2030 (and of 17–60 cm by 2090), seawater intrusion and coastal surges may be an increasingly frequent occurrence (Pacific Climate Change Science Program, 2011; The World Bank Group, 2011b).

3.4.4 Health risks related to climate change

The major health issues related to climate change in Papua New Guinea were identified through a vulnerability analysis of health risks by the multisectoral national working group on climate change, supported by external consultants. Four priority health problems were identified, and a vulnerability assessment was conducted in four geographic regions in conjunction with a review of data regarding natural disasters and other pertinent indicators (WHO, 2010a).

- **Vector-borne diseases**

Malaria is a priority health issue in Papua New Guinea, which has the highest burden of the disease in the Western Pacific Region (WHO, 2012c). The disease is endemic in lowland areas, but transmission cannot be sustained in cooler highland villages. However, the highlands experience malaria outbreaks following the periodic introduction of parasites to highland communities. A warming climate and other factors may facilitate expansion of sustained transmission of malaria at higher altitudes, and this is a commonly cited climate change and health risk in Papua New Guinea. Epidemiological data on dengue and other endemic vector-borne diseases, such as lymphatic filariasis and chikungunya, are relatively sparse but as in other settings, increased incidence of these diseases in wetter and warmer conditions is a risk. There are many dengue cases reported from Australian travellers coming back from

Papua New Guinea (Warrilow et al., 2012), but there is no reliable dengue surveillance in Papua New Guinea.

- **Water, extreme weather events and health**

Papua New Guinea is highly exposed to extreme weather events, but these phenomena are not new. However, an increase in weather variability is projected as a consequence of climate change. The frequency and intensity of events such as coastal erosion, cyclones, landslides and flooding are expected to increase with associated health impacts, including fatalities, injuries, malnutrition and psychosocial effects related to resettlement. These health effects are closely related to impacts of water and sanitation: less than 20% of Papua New Guinea's population has access to piped or well water in their homes and over 60% use spring, river or stream water. The primary climate change risks relating to water include issues of polluted water supplies, flooding, scarcity of freshwater and subsequent effects, including diarrhoeal diseases. Data suggest diarrhoeal diseases are associated with rainfall, so climate change impacts may affect incidence. Cholera was first reported in 2009 in Papua New Guinea, and it eventually spread along the coast and to the remote islands, although it has not been reported since 2011.

- **Malnutrition, and maternal and child health**

Increased climate variability reduces the predictability of harvests and has adverse impacts on food security. This is an especially pertinent issue in a largely agrarian country such as Papua New Guinea, where health impacts including malnutrition, low birth weight and paediatric developmental delays are prominent risks. Fishing yields may be affected by adverse weather and ecosystem destruction, and increases in the incidence of ciguatera fish poisoning are likely due to increased sea-surface temperature and reef disturbance, resulting in changes to diet, reduced protein intake and associated sequelae (Pacific Climate Change Science Program, 2011). Malnutrition in the highlands reflects a sociocultural situation that limits the access of children and women to food during drought months. Malnutrition is both one of the determinants and consequences of the problem and is associated with other health issues such as pneumonia, malaria, stunting and low birth weight (National Department of Health, Papua New Guinea, 2011).

3.4.5 Vulnerability assessment

Vulnerable population groups are those with the least adaptive capacity and include children, older people, those with underlying illnesses or disabilities, and economically disadvantaged populations. Papua New Guinea is prone to natural disasters including cyclones, flooding, landslides, drought and saltwater intrusion, making almost the entire population vulnerable to the impacts of changing climate patterns and frequency of extreme weather events. While rural and isolated populations are vulnerable due to their dependence on the environment, they have a history of coping mechanisms to address adversity and unemployment. Semi-urban groups are also highly vulnerable (Commonwealth Health Ministers' Update, 2009; WHO, 2010a).

3.4.6 Governance and national activity on climate change and health

In 2012, the Prime Minister of Papua New Guinea appointed a Minister of Climate Change, underlining the importance the Government attaches to the climate agenda. The Office of Climate Change and Development (OCCD) is the coordinating government body for climate change-related policies and actions in Papua New Guinea and works at the national level on research, analysis, and the development of the policy and legislative framework for the management of climate change. OCCD also works in consultation with other agencies and oversees the work of four cross-departmental working groups that are populated by representatives of civil society and the private sector (Fig. 18). However, the National Department of Health climate change and health strategies have yet to be integrated into national climate change mitigation and adaptation plans. Papua New Guinea has developed a framework for a long-term strategy, *Papua New Guinea Vision 2050*, which articulates the country's development initiatives. One pillar of the vision is environmental sustainability and climate change, and it is hoped health will be incorporated into these plans (Asian Development Bank, 2012; Office of Climate Change and Development Government of Papua New Guinea, 2013).

3.4.7 Health adaptation activities

Climate change and health adaptation have been constrained by overall health sector limitations associated with infrastructure, logistics and a shortage of trained medical personnel. Mainstreaming climate change adaptation into the health sector simply has not been a priority. As part of addressing broader disaster preparedness, which is highly relevant to climate change adaptation, the Government of Papua New Guinea adopted a disaster preparedness plan for health facilities (Government of Papua New Guinea, 2010). The National Disaster Committee provides considerable support to this effort, promoting

Figure 18. Schematic relationship of the climate change governance arrangement in Papua New Guinea

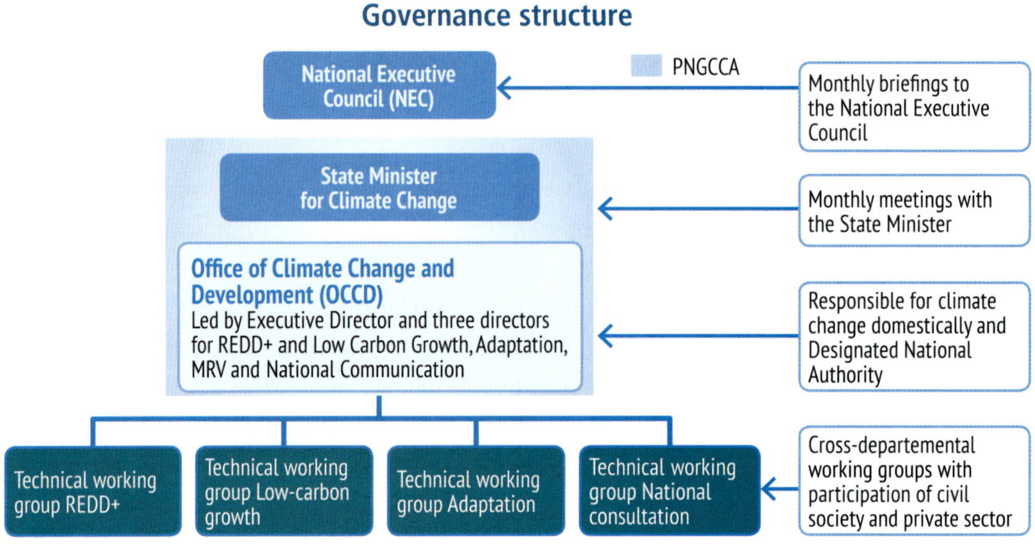

Note: REDD+ = Reducing Emissions from Deforestation and Forest Degradation

Source: Office of Climate Change and Development Government of Papua New Guinea, 2013.

awareness and preparedness for natural and other hazards throughout the country as well as providing technical advice relevant to natural and other hazards. To communicate during emergencies, the National Disaster Centre has a nationwide radio communication network that links provincial disaster response offices with the central level and also broadcasts via the National Department of Health radio network (National Disaster Centre, Papua New Guinea, 2005).

To address a specific issue identified during the national vulnerability analysis, a climate change and adaptation project was initiated in 2011 on the risk of malaria and other vector-borne disease outbreaks in the highlands. A technical working group of representatives from environment and other sectors was convened to ensure the project was in line with ongoing initiatives. The project aimed to build capacity to monitor the spread of diseases and vectors, provide training on diagnosis and treatment to medical staff, and communicate risks and protective behaviour to enable communities and health professionals to be aware of – and respond to – identified threats. Continual monitoring of epidemiological, vector and climatic parameters at the project site will generate further information about climate change and health impacts in Papua New Guinea (WHO, 2012b).

The WHO Regional Office for the Western Pacific supported a programme called Climate Change, Water and Health in Papua New Guinea in an effort to provide knowledge and information on climate change and water and health issues; to facilitate the process of planning and implementation; to provide advice on water quality monitoring; and to conduct hands-on training with the support of the Ministry of Environment, Republic of Korea. Water-related issues in Papua New Guinea were identified and listed, a work plan and teamwork for the project were outlined in a workshop, and water quality was assessed through site visits. Sanitation and maintenance of the rural water supply needed improvement and better management. Water source contamination in relation to mining is also an issue. A nationwide survey on water quality is needed to measure organic and inorganic materials, including metals. National and regional strategies to provide safe water should be established. Designing and installing sustainable and affordable forms of sanitation for private and public facilities is a priority.

3.5 Philippines

3.5.1 Geography, population and health status

The Philippines is a mountainous archipelago of over 7000 islands located in the western Pacific Ocean. The population at the 2007 census was 89 million people, living in 80 administratively decentralized provinces. The country is extremely diverse, with 180 ethnic groups, though 50.3% of the population lives in urban areas. While much of the urban population lives in slums and shantytowns, both urban and rural poverty are declining. The population is young: 33.8% are 14 years or younger and only 4.4% are over 65 years. Rapid urbanization, particularly in Metropolitan Manila, continues to create problems in housing, road traffic, pollution and crime (Asia Pacific Observatory on Health Systems and Policies, 2011; WHO & Department of Health, Philippines, 2012; WHO, 2011b).

NCDs account for six of the top 10 causes of death in the Philippines, with those of the heart and vascular system, malignant neoplasms, diabetes and chronic lower-respiratory diseases causing most deaths. Communicable diseases remain a significant source of morbidity, and of these pneumonia and tuberculosis continue to cause a significant number of deaths. Accidents of all types are ranked 10th among the causes of mortality. Maternal mortality is declining, but at a rate inadequate to achieve stated targets. Located along the "Pacific ring of fire" and the typhoon belt, the country has consistently been among those most affected worldwide by natural disasters. In 2009, the Philippines had the third highest number of deaths (1334) and second highest number of victims (13.4 million) from natural disasters (WHO, 2011b).

3.5.2 Greenhouse gas emissions

Globally, the Philippines is a minor emitter of GHGs, but cost-effective mitigation presents opportunities that should be captured as the country is a signatory to the UNFCCC and its Kyoto Protocol. The country accounted for less than 0.3% of global GHG emissions in 2008. However, emissions are on the rise. These are dominated by the energy and agricultural sectors.

3.5.3 Future climate projections

The Philippines has a humid equatorial climate marked by high temperatures and heavy rainfall. Annual rainfall measures as much as 5000 mm in the mountainous parts of the country, but less than 1000 mm in some sheltered valleys. The mean annual temperature is approximately 27 °C, with a temperature peak from April to June. Because of its location in the Western Pacific typhoon belt, the Philippines experiences an average of 20 typhoons each year (Yusuf & Francisco, 2010). However, climate patterns are not uniform across the country (Asia Pacific Observatory on Health Systems and Policies, 2011; Government of the Philippines, 1999).

Increasing temperatures over the past 60 years have been observed in the Philippines: from 1951 to 2010, the country experienced an average increase of 0.01 °C per year with the maximum and minimum temperatures increasing by 0.36 °C and 1.0 °C, respectively. There has also been an increase in the number of hot days and a decrease in number of cool nights, changes that are unprecedented over the past 140 years. In the future, all areas of the Philippines are predicted to warm; annual mean temperatures are expected to rise by 0.9 °C to 1.1 °C by 2020 and by 1.8 °C to 2.2 °C by 2050. Rainfall patterns are expected to change, and in most provinces the dry season will become drier and the monsoon season wetter. On the southern island of Mindanao, a reduction in rainfall is projected by 2050. Considerable variability remains between emission scenarios, but there is a trend to more extreme weather events (Regional Climate Change Adaptation Knowledge Platform for Asia, 2012).

3.5.4 Health risks related to climate change

The Philippines is a geographically, socially and ethnically diverse country, and these factors are reflected in climate change planning as well as health vulnerability and adaptation planning. The most substantial work in the health sector to date is a climate change vulnerability and impact assessment conducted by the Institute of Health Policy and Development Studies, National Institutes of Health, University of the Philippines, which will be included in operational guidelines of the Department of Health (National Institutes of Health, 2011). Vulnerabilities were examined according to the following categories: individual, family and community; socioeconomic factors; pathogen factors; health system and infrastructure; and national and local policy development and environmental policy. The assessment used a number of methods including literature review, epidemiological modelling and vulnerability mapping. Different adaptation options were costed according to health outcomes.

The Philippines is frequently affected by Pacific typhoons, causing a high number of fatalities, injuries, population displacement and loss of earnings that lead to secondary health impacts. Any increase in the frequency of these events as a consequence of climate change will impact health, with substantial vulnerable populations.

The altered epidemiology of communicable diseases due to the seasonal patterns and climatic associations of dengue, malaria and waterborne diseases is a prominent risk. The mechanisms of changing incidence may be different from other countries. In the Philippines, vector mosquitoes for disease are already ubiquitous and increased temperature or humidity is unlikely to increase their density or range. However, risks remain of increased introduction of pathogens and serotypes from other countries whose disease burdens are impacted by a changing climate, or of an increasing season of transmission of climate-sensitive diseases (Manila Observatory, 2010). An increased number of cases of the waterborne disease leptospirosis were highlighted as a particular concern due to frequent floods in many areas of the country and exposure from working in rice fields.

3.5.5 Vulnerability assessment

The Philippines is highly exposed to natural disasters, including typhoons (also known as cyclones), floods and associated events such as landslides. The most exposed is the Eastern Visayas region, in the centre of the archipelago, which faces the Pacific and is most frequently affected by the landfall of typhoons. However, it is important to note that the effects of climate change in the Philippines are unlikely to introduce new health issues or challenges to the country. Rather, the likely effects are an increase in events and health impacts that have been the target of government activity for some time. Foremost among these is disaster response, for example, in the Visayas region. Lifestyles will be less predictable and fishing catches, agricultural yields and the loss of livelihood will result in declining population health. Declines in scarce groundwater may have particularly adverse effects, placing additional pressure on sanitation systems and resulting in increased incidence of diseases (Manila Observatory, 2010).

3.5.6 Governance and national activity on climate change and health

The Philippine Government considers climate change a priority and has accordingly developed a number of policy measures and strategies to address challenges. The climate change law was enacted in October 2009 and mandated the creation of an autonomous interagency body, the Climate Change Commission (CCC) (Fig. 19). CCC is a part of the Office of the President and develops policy and coordinates, monitors and evaluates the programmes and action plans of the Government relating to climate change. It also facilitates mainstreaming of climate change concerns into government plans and actions and serves as a coordinating mechanism among government agencies on climate change activities. The climate change

Figure 19. Climate Change Commission organizational chart, Philippines

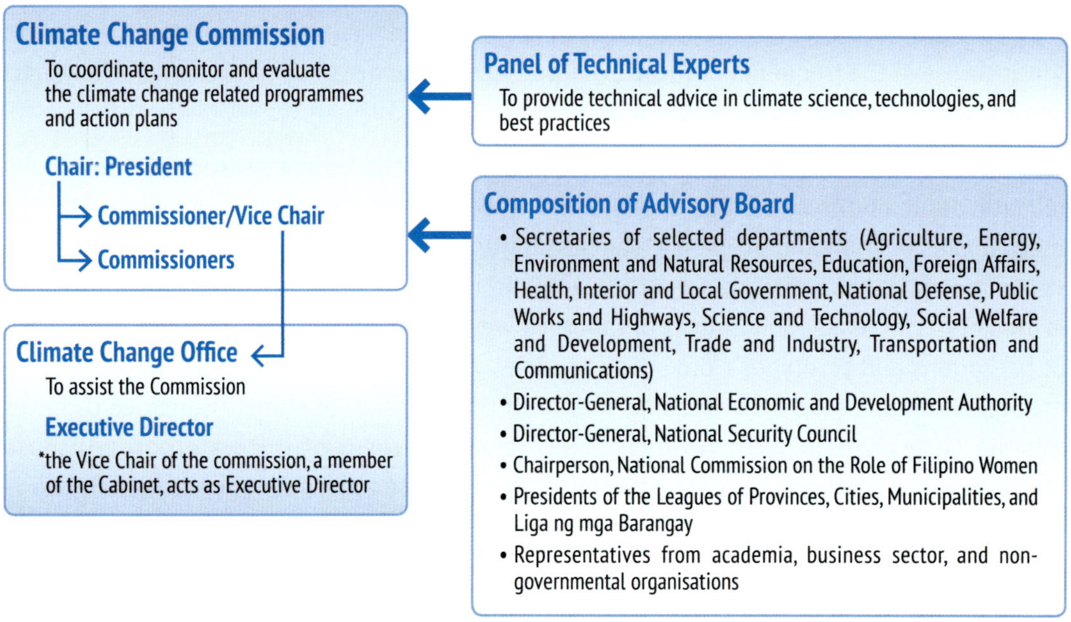

Source: Republic Act No. 9729, Republic of the Philippines.

law additionally directed the development of a *National Framework Strategy on Climate Change 2010–2022* (NFSCC), guiding both mitigation and adaptation efforts in the country (Climate Change Commission, Office of the President of the Philippines, 2010).

The National Policy on Climate Change Adaptation for the Health Sector was signed by the Secretary of the Department of Health in March 2012 and guides overall health sector climate change and health implementation. The policy encourages mainstreaming of climate change activities into ongoing Department of Health programmes and will increase the capacity of other health programmes and local government units to manage climate change impacts (Department of Health Philippines & WHO, 2012).

3.5.7 Health adaptation activities

With a strong policy platform articulated under the NFSCC, adaptation activities are under way in many sectors in the Philippines. The *National Climate Change Action Plan* (NCCAP) was approved by the Government of the Philippines in April 2010 and describes seven strategic priorities:

- food security,
- water sufficiency,
- environmental and ecological stability,
- human security,
- sustainable energy,
- climate smart industries and services, and
- knowledge and capacity development.

Three of these priorities are closely related to human health: food security, water sufficiency and human security (Climate Change Commission, Office of the President of the Philippines, 2011). Specifically within the Department of Health, policies and a strategic plan have been developed that support health sector strengthening to enhance responses to the impact of climate change and a Climate Change Unit has been created to articulate these policies. Surveillance and response capacities have been improved to strengthen responses to disasters and a train-the-trainers exercise has been conducted to build capacity on climate change and health for health workers. A risk communication plan has been developed and information, education and communication materials for health promotion are available (Regional Climate Change Adaptation Knowledge Platform for Asia, 2012).

Box 2 Best practices: local-level assessment of climate change and health vulnerability

An assessment on the Impact of Climate Change on Health was conducted for the Department of Health, the National Economic and Development Authority and WHO. It revealed that many cities in Metropolitan Manila are vulnerable to vector-borne and waterborne diseases, such as dengue, typhoid, cholera and leptospirosis. To identify specific vulnerabilities at the local level, the distribution of climate-sensitive diseases and their relation with climate variability were examined in four cities and a municipality in Metropolitan Manila. Highest vulnerability was found in populations of the following cities and their level of adaptation capacity identified:

Pateros – level of adaptive capacity to dengue, typhoid, cholera and leptospirosis is high.

Pasig – level of adaptive capacity to the four diseases is high.

Taguig – level of adaptive capacity to dengue and leptospirosis is high, but to typhoid and cholera is medium.

Marikina – level of adaptive capacity to dengue is high, to typhoid and leptospirosis is medium, and to cholera is low.

Quezon City – level of adaptive capacity to all four diseases is high.

A grade of high is given if the health system is well-equipped to respond to the disease and has the ability to prevent transmission. A grade of medium is given if the health system is equipped with the basic features in the three key areas of technology, information and skills, and institutions (gaps could include lack of equipment and/or computerized systems for reporting). A low grade is given if the capacity is lacking in all the parameters of technology, information and skills, and institutions (additional gaps could include lack of trained personnel, no information available and/or no policies in place).

The results of the study served as the basis for the formulation and adoption of local-level policies and actions programmes to reduce the impacts of climate change in spreading the diseases, inform and educate residents, improve health-care systems, and strengthen the capacity of health centres to respond to emergency situations.

Source: modified from Department of Health Philippines & WHO, 2012.

Box 3 Best practice example: Surveillance in Post-Extreme Emergencies and Disasters (SPEED)

SPEED is a surveillance system developed to provide real-time health information reporting after a disaster. It assists public health managers during disease outbreaks due to disasters (Fig. 20). It was developed and implemented by the Philippine Department of Health and WHO, with funding support from the Australian Department of Foreign Affairs and Trade, the United States Agency for International Development (USAID), the Government of Finland and the European Union.

It was developed in response to three consecutive powerful tropical storms (Ketsana, Pharma and Santi) in 2009, which caused more than 200 deaths and 3000 cases of various diseases. The initiative requires a SPEED reporter who could either be a barangay health worker, midwife, nurse or physician tasked to submit a daily report for SPEED and a physician who manages the SPEED reports in the health facility. SPEED operates under three phases:

1. the submission of data where a SPEED form is first completed to serve as a backup prior to submitting the report through a text-messaging system;
2. validation of data by the medical health officer/physician by logging on to the SPEED website to validate the information either through mobile or online access; and
3. generation of reports at the municipal, provincial, regional or national levels.

It uses syndromic surveillance and adopts available information and communication technology, such as text messaging and the Internet for data collection, analysis and report generation. It focuses on 21 of the most common health conditions encountered after disasters. It is not a substitute but rather complements the existing routine surveillance system in emergency situations.

This programme worked in the response stage of the unprecedented impact of Typhoon Haiyan (Yolanda in Philippines) in Tacloban, Leyte Province (Department of Health, Philippines, 2009).

Figure 20. Brochure for SPEED

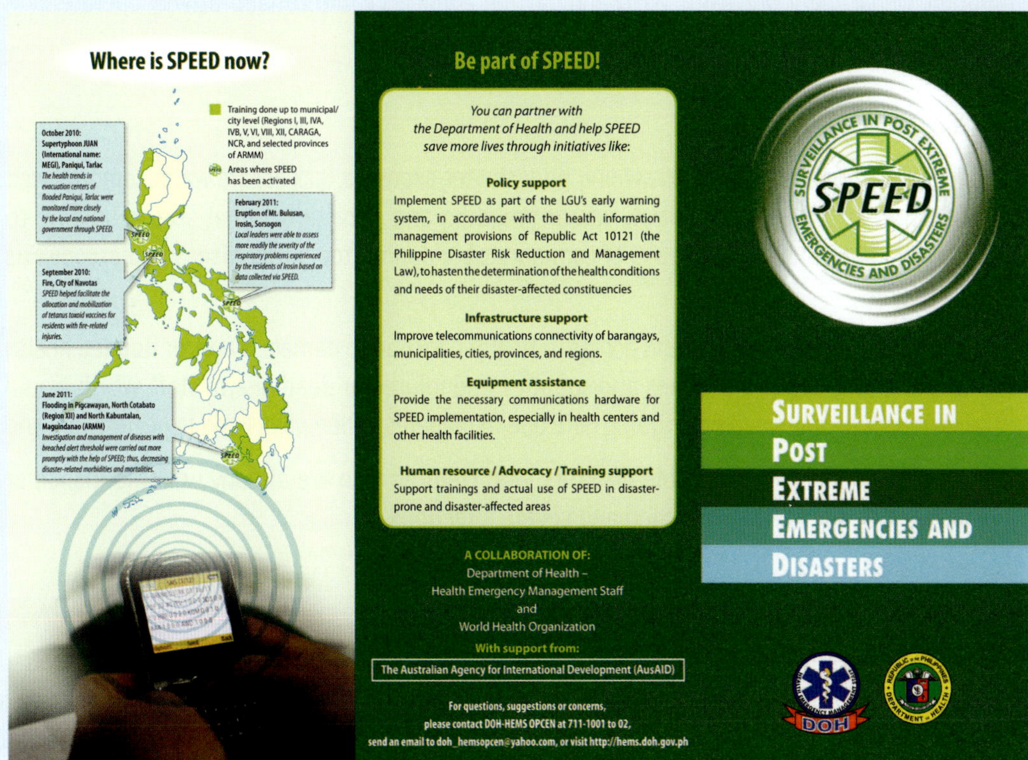

Source: Department of Health, Philippines & Office of the WHO Representative in the Philippines, 2011.

Box 4 Supertyphoon Haiyan

On 8 November 2013 at 04:40 a category 5 typhoon, Haiyan (locally named Yolanda), made landfall in Guiuan municipality, Eastern Samar province, moving steadily north onto Cebu island, with maximum winds of 235 km/h and severe gusts of 275 km/h. The typhoon made subsequent landfalls in Tolosa municipality, south of Tacloban City, Leyte province, Daanbantayan and Bantayan island, Cebu province, and Conception, Iloilo province. The typhoon affected the city of Roxas in Capiz province and the tourism centre of Boracay in Aklan province.

Box 4 Supertyphoon Haiyan (continued)

At least 18 million people were living in the worst-affected regions.

The Secretary of Health formally requested assistance from WHO and partners, and the United Nations declared a Level 3 emergency.

By 6 December, the National Disaster Risk Reduction Management Council of the Philippines reported 5786 deaths, 26 233 injured and 1786 missing, with over 11 million people affected by the storm, including over 4 million displaced people of whom 94 310 were in 385 evacuation centres. In these areas, medical personnel were heavily affected and health facilities were extensively damaged, including submerged ground floors and damaged and destroyed medical supplies, equipment, records, office equipment and buildings.

On 16 November, WHO identified the major public health risks given the underlying demographic and disease profile and context of the Philippines, as well as the dangers presented given the typhoon and its aftermath. The identification of the risks provided a framework by which the Philippine Department of Health, WHO and partner agencies could collaboratively operate.

The major risks to public health identified were:

- major trauma and injuries;
- population displacement, overcrowding, poor shelter, exposure, lack of safe water, compromised sanitation and hygiene facilities, vector breeding and poor nutritional status, leading to increased potential for outbreaks of communicable and vector-borne diseases, leptospirosis and tetanus;
- disruption of health services and access to health care due to damaged and/or flooded health infrastructure, loss of medicines and supplies, and injured or dead health staff, which would also impact people needing ongoing care due to diabetes, tuberculosis and hypertension; and
- malnutrition and communicable diseases, with children who are already undernourished or malnourished prone to prolonged and severe infections.

Source: WHO, 2013.

At the national level, the assessment described climate change health risks that should be considered and identified four priority areas for health sector adaptation: 1) disease and vector control; 2) land use and habitation; 3) water and sanitation; and 4) solid waste management. Specific activities include achieving improvements in hygiene and sanitation, the installation of disinfection facilities and strengthening waste disposal. In disaster risk reduction, adaptation activities focus on reducing risks through strengthening structures and introducing incentives and payments to allow those affected to relocate. Different options should be examined for cost-effectiveness where possible. Adaptation to possible increases in the malaria burden would require boosting microscopy capacity, liaison with disaster coordinating bodies and working with indigenous populations. The use of insecticidal bednets was identified as the intervention that would help reduce disability-adjusted life years (DALYs), but early diagnosis and treatment are the most cost-effective approaches at the community level.

From a health systems approach and also as part of an effort to develop evidence of impacts, surveillance systems should be strengthened, including vector surveillance. In addition, measures could be taken to improve disaster response and information dissemination, including epidemic forecasting and ensuring sufficient medical stockpiles are in place.

At the local level, the assessment provided step-by-step procedures and examples of vulnerability maps developed for different health outcomes, such as flooding, typhoons or communicable disease outbreaks. Local government units are encouraged to map local vulnerabilities and use these maps in local-level decision-making where technical and human resources are adequate. Local government units are also encouraged to take steps to provide effective diagnosis and treatment with a view to universal health care, and take a lead in forging intersectoral collaboration.

The Philippines has also developed a variety of education and awareness-raising materials relating to climate change and health impacts to inform the public of the risks of leptospirosis, cholera, dengue and measles (Fig. 21).

Figure 21. Materials developed by the national Department of Health and partners on the communicable disease risks of climate change

Source: Department of Health Philippines, 2013.

3.6 Republic of Korea[3]

3.6.1 Geography, population and health status

The Republic of Korea is located at the eastern edge of the Asian continent. The capital is Seoul, the total area is 100 210 km^2, and the per capita gross domestic product (GDP) is US$ 17 074 as of 2009. The population was 49.7 million people in 2009, an increase of approximately 50% since 1970. The average life expectancy in 2009 was 76.99 years for men and 83.77 years for women.

3.6.2 Greenhouse gas emissions

Since 1990, Republic of Korea's GHG emissions have been growing at 4.5% per year. Although the growth rate for total emissions from the energy and industrial process sectors slowed somewhat in 2008, they still posted a 2.9% rise from the previous year at 585.9 million tonnes CO_2e. For the energy sector, energy conversion (power generation) accounted for the largest portion, followed by the industrial, transportation, household and commercial sectors. In November 2009, the Republic of Korea set concrete national mid-term emissions reduction goals (a 30% cut from 2020 BAU). To carry this out, the Greenhouse Gases and Energy Target Management System for multiple emissions sources is being used according to the Enforcement Decree of the Framework Act on Low Carbon and Green Growth, and reduction goals are set and underway by stage (five-year units), sector and industry.

3.6.3 Future climate projections

The mean temperature in the Republic of Korea's six major cities has risen 1.7 – since 1912 and is predicted to increase a further 2 – by 2050 compared to 2000.

Annual precipitation for the country's six major cities – Busan, Daegu, Daejeon, Gwangju, Incheon and Seoul – rose 19% over the past 100 years. Although the number of days with precipitation decreased by 14%, rainfall intensity increased by 18%. Days with torrential rainfall of at least 80 mm have doubled compared to the 1970s. According to predictions of future precipitation on the Korean Peninsula, the number of such days will climb 15% by 2050 and 17% by 2100 compared to 2000. Also, it is expected that temporal and spatial variability

3. The information for this chapter has been taken entirely from *2010 Climate Change and Human Health: Impact and Adaptation Strategies in the Republic of Korea,* published by Korea Centers for Disease Control and Prevention.

will increase, that drought and torrential rainfall will intensify, and that precipitation will increase during August and September.

From 1963 to 2006, the sea level on the coast of the Korean Peninsula rose approximately 8 cm. In the Jeju region it rose by as much as 22 cm during the same period. It is expected to rise 9.5 cm by 2050 and 20.9 cm by 2100 compared with 2008. The surface temperature for neighbouring waters of the Korean Peninsula rose by an average of 1.31 – for the 41-year period ending in 2008, which is a figure far exceeding the worldwide mean rise of 0.5. According to predictions, future rises in seawater temperatures around the Korean Peninsula are expected to climb 1.3 – by 2050 and 2.9 – by 2100 compared to 2008.

3.6.4 Health risks related to climate change

The health effects of climate change are occurring with greater intensity than anticipated. The following is a summary of the current status of the impact on human health from climate change in the Republic of Korea and adaptation strategies currently being conducted at the Ministry of Health and Welfare and Korea Centers for Disease Control and Prevention, in terms of direct impacts of heatwaves and extreme weather events and indirect impacts of vector-borne diseases, waterborne and foodborne diseases, and pollen allergen air pollution.

- **Heatwaves**

The Republic of Korea is experiencing a rise in mean temperatures, along with an increase in the incidence of abnormally high temperatures. Maximum temperatures in Seoul steadily increased between 1971 and 2007. There were a great number of extreme heat days in 1994 and 1997, with the maximum daily temperatures of at least 30 – occurring 56 times and 61 times, respectively, in those years (Fig. 22).

Figure 22. Number of extreme heat days in Seoul, 1971–2007

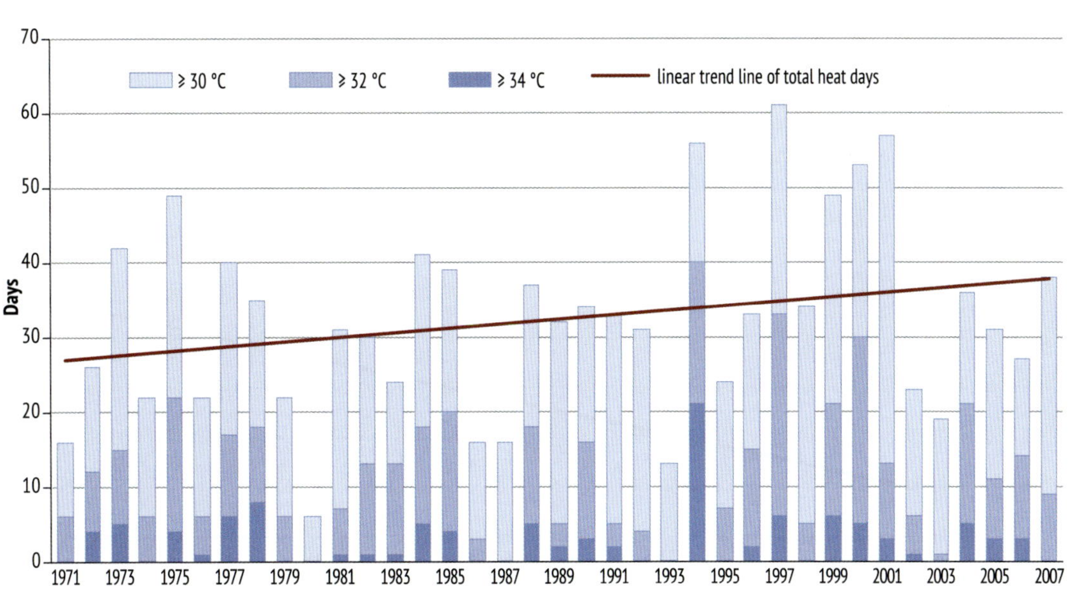

Source: Ministry of Health and Welfare, Republic of Korea, 2008.

According to an analysis of the relationship between the Republic of Korea's summer season (July and August) temperatures and mortality between 1991 and 2005, the relation between the mean value of the maximum daily temperatures and deaths was found to be statistically significant (Fig. 23).

Considering the trend of frequent summer season high temperatures, excess deaths associated with those high temperatures are very likely to keep rising. The extent of adaptation or sensitivity towards extreme heat and the threshold temperature at which the death toll abruptly rises differ depending on the region.

Extreme weather events

The incidence of extreme weather events in the Republic of Korea peaked in the latter half of 1980s and has subsequently declined. Nevertheless, the average duration of extreme weather events has been on the rise since mid-1990s, and such an event lasted for 11.3 days in 2005 (Fig. 24).

Vector-borne diseases

The incidence of infectious diseases in the Republic of Korea is steadily declining as a result of improved hygiene, vaccination and strengthening of the health-care system. However, since the 1990s, diseases classified as highly related to climate change, such as scrub typhus,

Figure 23. Temperature in summer (July–August) and deaths in Seoul, 1991–2005

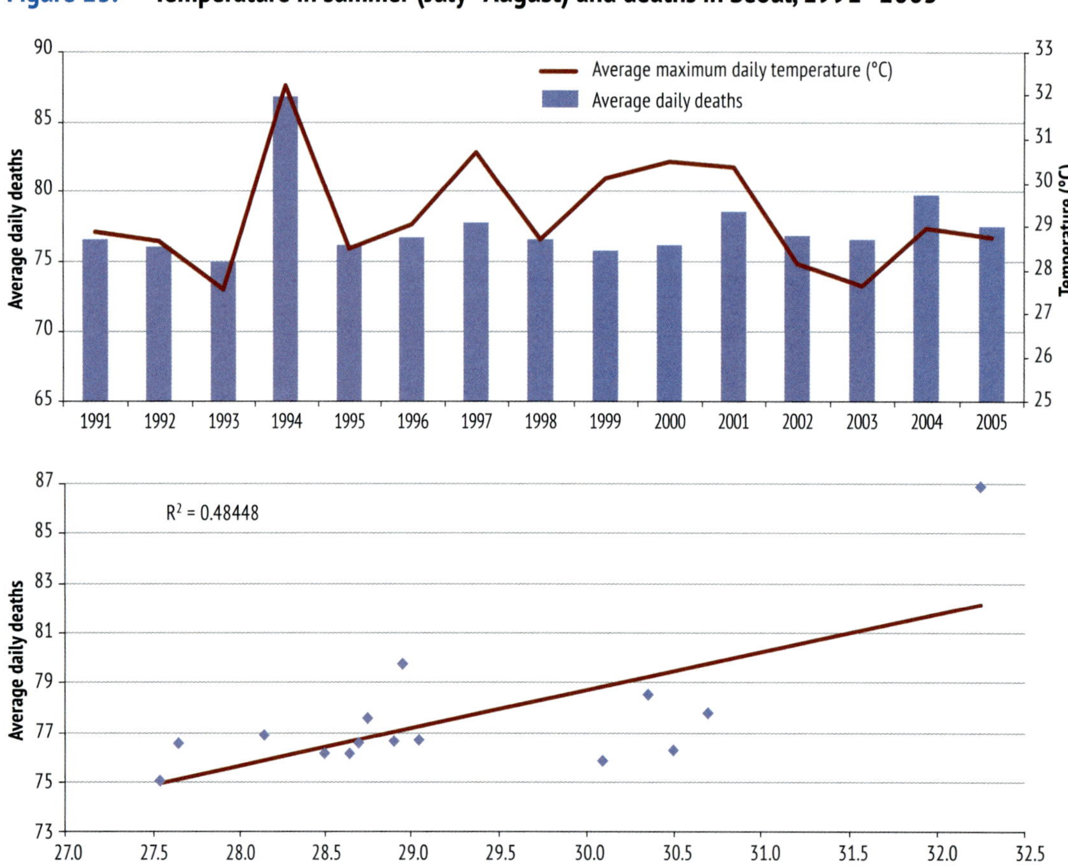

Source: Ministry of Health and Welfare, Republic of Korea, 2008.

malaria, shigellosis, haemorrhagic fever with renal syndrome, leptospirosis, murine typhus and similar diseases, are on the rise.

Malaria re-emerged in 1993, and its incidence increased slowly between 1994 and 1996, then sharply rose in 1998 and peaked in 2000. It gradually decreased until 2004 and has fluctuated since then. Rodent-borne infectious diseases, such as scrub typhus, haemorrhagic fever with renal syndrome, leptospirosis and similar diseases, are all on the rise (Fig. 25).

Figure 24. Disaster incidence by year and average disaster duration, 1981–2006

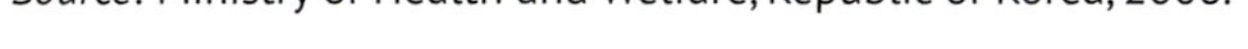

Source: Ministry of Health and Welfare, Republic of Korea, 2008.

Figure 25. Yearly incidence of major vector-borne diseases, 1997–2009

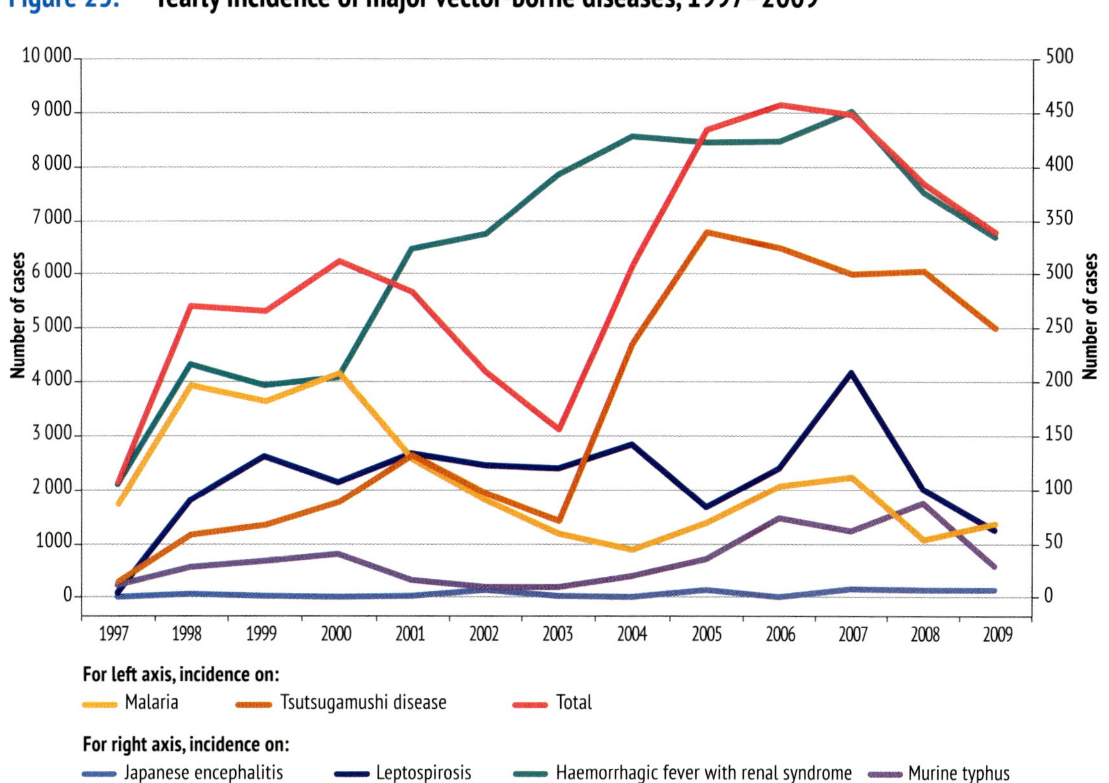

Source: Korea Centers for Disease Control and Prevention, 2010.

Waterborne and foodborne diseases

Between 1968 and 2006, the sea surface temperature of the waters around the Korean Peninsula increased by an average of 0.93 °C. It increased by 0.80 °C in the East Sea, 1.04 °C in the South Sea, and 0.97 °C in West Sea. The rise in seawater temperature causes a proliferation of microbes such as vibrio, which may result in increased infectious diseases through seawater and seafood. Analysis of the relationship between coastal seawater temperatures and vibrio detection in some parts of Busan revealed that a high level of correlation was observed.

According to analysis of the prevalence of waterborne and foodborne diseases between 2007 and 2009, incidence between June and September accounted for 44.2% of total incidence and 45.2% of people with symptoms (Fig. 26).

Pollen, allergens and air pollution

Pollen, a key allergen, is affected by climatic factors. The higher the spring temperature, the faster the flowering period of tree pollen, lengthening the exposure time to pollen and at the same time increasing the total exposure volume. It is known that the greater the amount of pollen, the more severe the symptoms of allergy patients become, thus increasing the frequency of complaints of serious symptoms.

According to an analysis of patients who visited selected hospitals of Gyeonggi-do province between 1999 and 2000 and between 2002 and 2008, if the minimum temperature in March rises 1 °C, the probability that patients sensitized to tree pollen visit hospitals was shown to increase 12.1%. This seems to be due to the fact that, when the minimum temperature in March is high, the amount of tree pollen increases, thus making the symptoms of allergy patients more severe, driving up hospital visits (Fig. 27).

According to a comparison of air pollutant concentrations during the week of 22 July 1994 – the most severe heatwave period in Seoul in an average year – there was little variation

Figure 26. Monthly incidence of waterborne and foodborne diseases, 2007–2009

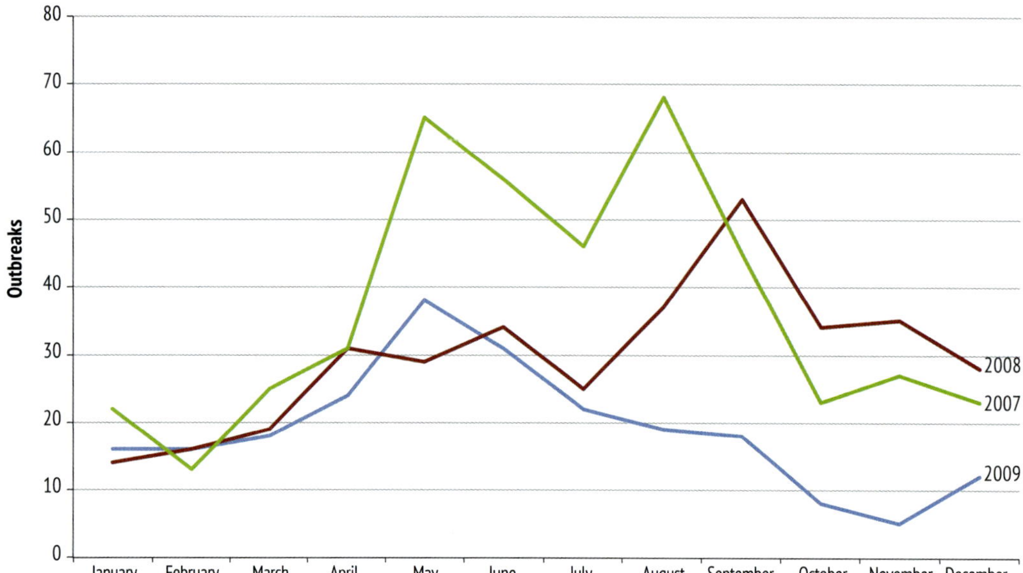

Source: Korea Centers for Disease Control and Prevention, 2010.

in the concentration of pollutants such as nitrogen dioxide and sulfurous acid gas. However, ozone levels were 63.8 parts per billion (ppb) in 1994, more or less a doubling from the average concentration of 34.1 ppb during the same week during the previous three years. The rise in temperature during the summer increases atmospheric ozone pollution, and thus may increase its impact on human health.

Respiratory allergic diseases such as asthma, allergic rhinitis, atopy and similar diseases are rapidly increasing both locally and abroad due to air pollution and ecological changes caused by climate change (Fig. 28). Accordingly, as individual and national medical expenses increase, there is urgent need for a response system.

Figure 27. Relationship between the minimum temperature in March and patients sensitized to tree pollen from April to July

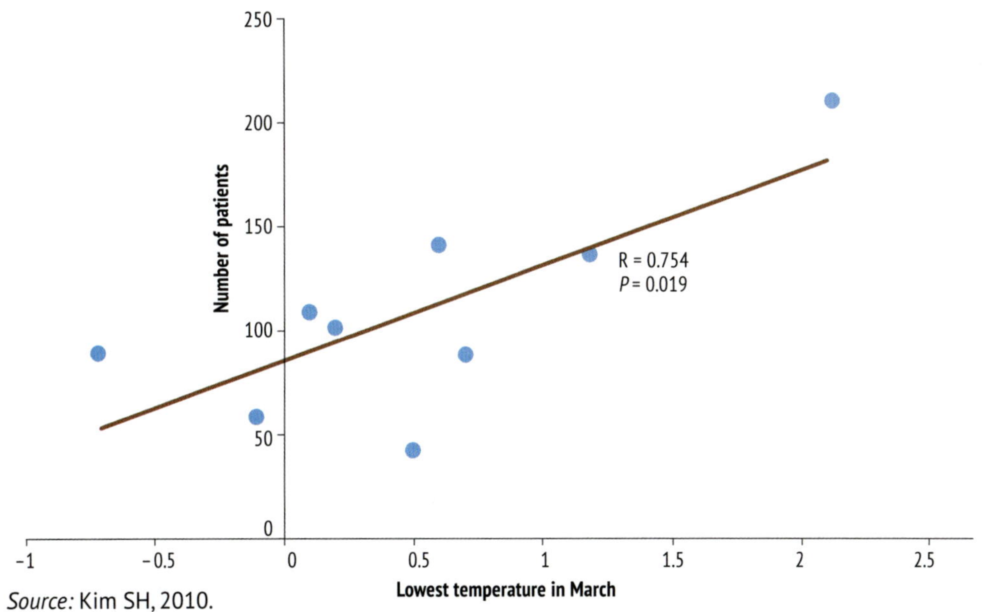

Source: Kim SH, 2010.

Figure 28. Increase in patients with allergic disease

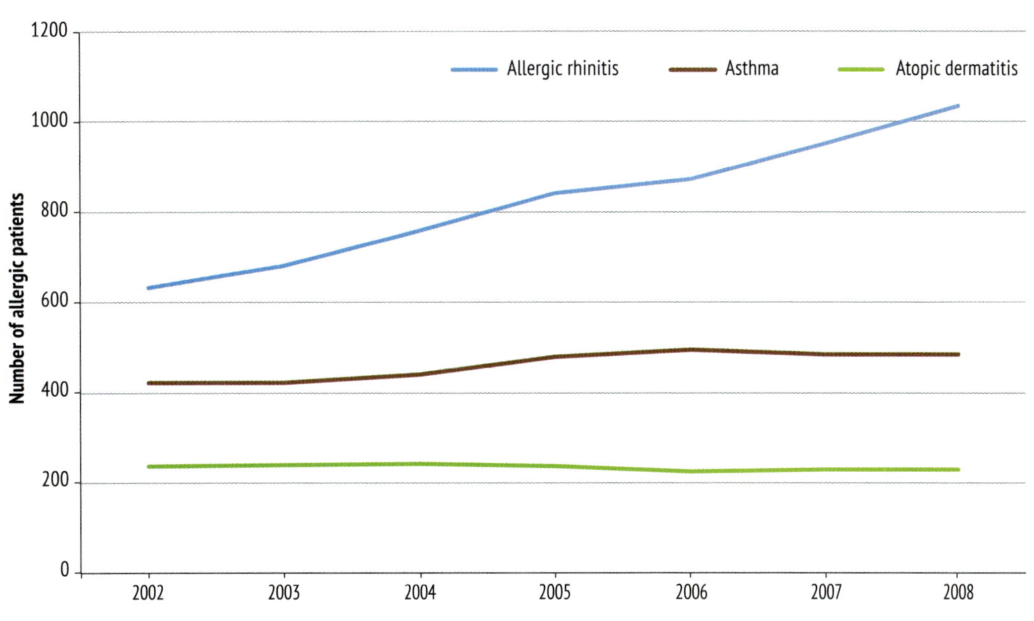

Source: Korea Centers for Disease Control and Prevention, 2010.

3.6.5 Vulnerability assessment

According to an analysis, by region, of deaths caused by extreme weather events, mortality is higher in small and medium-sized cities and in farming and fishing villages than in metropolitan areas. Coastal regions also show higher mortality than inland regions. Hence, people residing in coastal regions, in small- and medium-sized cities and in farming and fishing villages are 11 times more exposed to risk than those in the inland regions or metropolitan areas.

Malaria commonly afflicts soldiers stationed in the Gangwon, Gyeonggi and Incheon regions bordering the Democratic People's Republic of Korea, as well as residents around these regions. Rodent-borne diseases such as scrub typhus, haemorrhagic fever with renal syndrome and leptospirosis occur mostly during the autumn season, especially in October and November among farmers in rural areas such as Chungnam and Jeonbuk (Table 6).

Foodborne and waterborne diseases that were previously prevalent mainly during summer are increasing throughout the year due to school meals, climate change, transportation development, increased overseas travel and eating outside the home. *Vibrio vulnificus*, a legally designated and controlled infectious disease, mainly occurs during August and September. According to an analysis targeting patients between 1991 and 2007, the populations at risk were identified as those 65 years or older, those residing in the Gyeongnam and Jeonnam regions, the unemployed, and those involved in agriculture and fisheries. For shigellosis, the vulnerable populations were identified as students 14 years and younger and residents of Gyeongnam and Jeonnam regions.

Table 6. Groups vulnerable to vector-borne and waterborne diseases

	Gender	Age	Region	Occupation	Period
Vector-borne diseases					
Malaria	Men	20–64 years	Gyeonggi, Incheon, Gangwon	Military servicemen, Students	Jul.–Aug.
Scrub typhus	Women	65 years or more	Jeonbuk, Chungnam, Gyeonggi	Farmers and fishermen	Oct.–Nov.
Haemorrhagic fever with renal syndrome	Men	65 years or more	Jeonbuk, Chungnam	Farmers and fishermen	Oct.–Nov.
Leptospirosis	Men	65 years or more	Jeonnam, Jeonbuk	Farmers and fishermen	Oct.–Nov.
Waterborne diseases					
Vibrio vulnificus	Men	65 years or more	Jeonnam, Gyeongnam	Unemployed, farmers and fishermen	Aug.–Sept.
Shigellosis	Women	0–14 years	Jeonnam, Gyeongnam	Students	–

Source: Ministry of Health and Welfare, Republic of Korea, 2008.

3.6.6 Governance and national activity on climate change and health

A *Comprehensive Plan on National Adaptation to Climate Change* (2008) has been developed and updated, becoming the first national legal adaptation strategy as a result of enforcement of the Framework Act on Low Carbon and Green Growth. This is the blueprint for establishing detailed enforcement plans for central and provincial governments, and represents a next phase of development of adaptation plans in the National Strategies on Green Growth (2009). Considering the uncertainty of climate change impact, five-year integration plans will be established. The situation is to be monitored every year, and modifications and updating will reflect assessment results. Figure 29 shows the joint plan involving 13 central ministries and offices involved in 10 areas, including health and disaster.

3.6.7 Health adaptation activities

In April 2010, the Ministry of Health and Welfare established the *National Climate Change and Health Adaptation Action Plan (2010–2014)* based on the policy research task of developing a climate change impact monitoring system and adaptation strategies for human health. Based on this action plan, adaptation strategies are being established for the health sector. Modifications and updates are under way to establish a detailed implementation plan for the health sector under the title, National Climate Change Adaptation Strategies (2011–2015). There are six initiatives: 1) consolidation of infectious disease prevention and control to address climate change; 2) health management of vulnerable groups for air pollution; 3) health management of vulnerable groups for heatwaves; 4) full preparation for extreme weather events; 5) consolidation of research and developments to address climate change; and 6) formation of an adaptation basis for climate change (Fig. 30).

Figure 29. Areas and relevant agencies for National Adaptation Strategies for Climate Change

National Adaptation Strategies for Climate Change, 2011–2015
(led by the Ministry of Environment)

Health	Disaster	Agriculture	Forest	Ocean/ Fisheries	Water Management	Ecology	Surveillance & Prediction of Climate Change	Adaptation Industries & Energy	Education/ PR & International Cooperation
Ministry of Health & Welfare Ministry of Environment	Ministry of Public Administration & Security Ministry of Land, Transport & Maritime Affairs National Emergency Management Agency Ministry of Environment	Ministry for Food, Agriculture, Forestry & Fisheries Rural Development Administration	Forest Service	Ministry of Land, Transport & Maritime Affairs Ministry for Food, Agriculture, Forestry & Fisheries	Ministry of Land, Transport & Maritime Affairs Ministry of Environment	Ministry of Environment Ministry for Food, Agriculture, Forestry & Fisheries Ministry of Land, Transport & Maritime Affairs	Ministry of Environment Ministry of Education Science & Technology Meteorological Administration	Ministry of Knowledge Economy Ministry of Environment Ministry of Land, Transport & Maritime Affairs	Relevant government agencies

Source: Ministry of Health and Welfare, Republic of Korea, 2010.

Figure 30. Six initiatives to address climate change

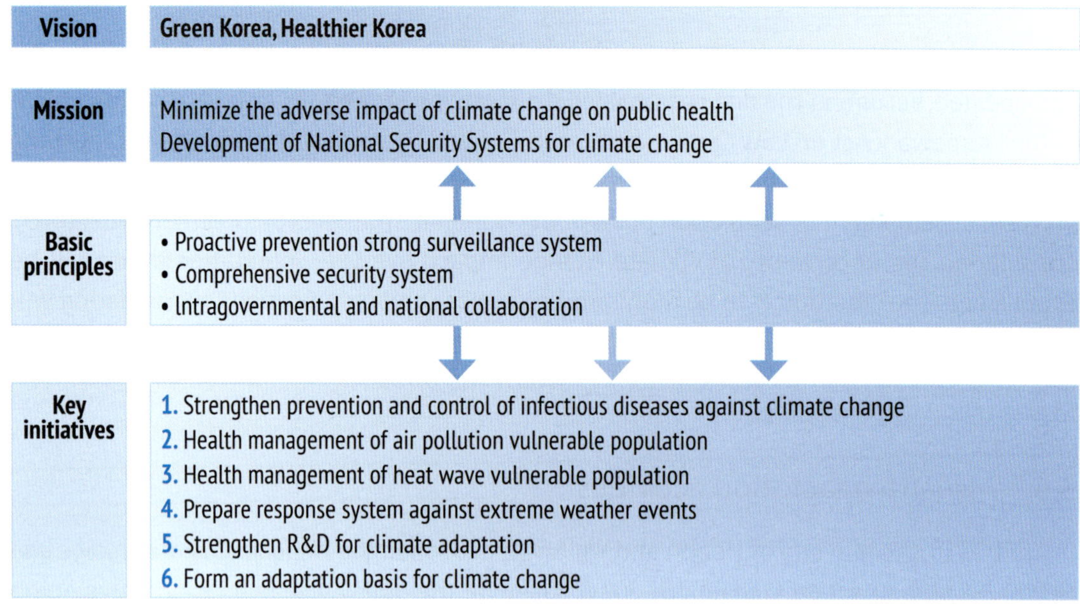

Source: Ministry of Health and Welfare, Republic of Korea, 2010.

Viet Nam

> **"Viet Nam is believed to be one of the most vulnerable countries to climate change in the world due to its long low-lying coastline and exposure to typhoons, storms, and heavy and variable rainfall.**
>
> Danish International Development Agency (DANIDA), Ministry of Foreign Affairs of Denmark, 2013

3.7.1 Geography, population and health status

Viet Nam is the easternmost country on the Indochina Peninsula, and is geographically diverse, consisting of mountains, plateaus, and estuarine and offshore islands, with a coastline of 3260 km. As of 2009, Viet Nam has a predominantly rural population of 86 million, with 54 distinct ethnic groups, many of whom live in mountainous and remote areas. There is a substantial degree of rural-to-urban migration. The country shares land borders with Cambodia, China and the Lao People's Democratic Republic and is an important maritime transit route between the Indian and Pacific oceans. Viet Nam's long coastline, geographic location, diverse topography and variable humid tropical to subtropical climate contribute to its being one of the most hazard-prone countries of the Asia-Pacific region, with storms and flooding, in particular, responsible for economic and human losses. Most typhoons occur between May and December. Given that a high proportion of the country's population and economic assets, including agriculture, are located in coastal lowlands and deltas, Viet Nam has been ranked among the five countries in the world most likely to be affected by climate change: a sea level rise of 1 m would directly affect 10% of the population by inundating their homes and a rise of 3 m would directly affect 25% of the population, with a loss of up to 25% of GDP (The World Bank Group, 2011c; WHO, 2011b).

Health status in Viet Nam is tightly linked to socioeconomic development, which has slowed following the global economic crisis, and is also linked to the effects of environmental and climate change (WHO & Ministry of Health Viet Nam, 2012). Life expectancy at birth was 70.2 for males and 75.6 for females in 2009, and since 1990 the maternal mortality ratio (MMR) has been falling steadily. However, additional efforts are required to achieve MMR goals. Historically, communicable diseases have caused most of the disease burden, but recently NCDs have been increasing in prominence. According to hospital data, leading causes of mortality include injuries, AIDS-related conditions, pneumonia, accidents and some NCDs. Between 1980 and 2009, natural disasters including typhoons, floods and droughts caused

15 917 deaths. Such disasters are becoming increasingly severe and frequent as a consequence of climate change (WHO, 2011b).

3.7.2 Greenhouse gas emissions

Viet Nam's most recent GHG inventory was conducted in 2000, when a total of 150.9 million tonnes CO_2e was recorded. Agriculture and energy were the industries with the highest level of emissions, representing 43.1% and 35.0% of the total, respectively. As the society industrializes, energy consumption is shifting towards industry and transport, and fossil fuels are increasingly important. From 1990 to 2007, CO_2 emissions per capita increased more than four-fold. While it is predicted that the land use and land use change sector will become a carbon "sink" due to strong forest management, the country as a whole will remain an emitter of CO_2 through 2030 (United Nations Viet Nam, 2013; Ministry of Natural Resources and Environment, Viet Nam, 2010).

3.7.3 Future climate projections

Viet Nam's climate varies from humid tropical to subtropical, and is characterized by strong monsoons, sunny days, high rainfall and high humidity. The geographical diversity is reflected in its climate, and the country is climatically divided into northern and southern areas by the Hai Van mountain pass. Northern areas experience cool winters with average temperatures of 15–19 °C and hot, wet summers which average above 30 °C. Southern areas are hot all year, with average temperatures of 25–27 °C. The long coastline is particularly vulnerable to Pacific cyclones and typhoons: on average, the country experiences four to five typhoons annually (Ministry of Natural Resources and Environment Socialist Republic of Viet Nam, 2003). Following typhoons and during the May–October monsoon season, flooding is common, including deluges of saline water. However, interannual variability in rainfall – often associated with the El Niño southern oscillation – can result in flood-prone areas also suffering from droughts during the dry season (The World Bank Group, 2011c).

Due to its climatic heterogeneity, climate change is projected to have different impacts in different parts of the country. Broad, national-level climate models predict increased mean annual temperatures of approximately 1 °C by 2050; these will be accompanied by an increase in the number of heatwaves and a reduction in the number of very cold periods. While a clear picture of future precipitation is complicated by model uncertainties, most models show increased precipitation. However, it seems clear that there will be increased climatic variability and increased frequency of extreme events. Sea level rises will particularly impact the low-lying southern delta region and are predicted to be between 28 cm and 33 cm by 2050 (The World Bank Group, 2011c; United Nations Development Programme, 2008).

3.7.4 Health risks related to climate change

Approximately 80% of Viet Nam's population uses water from wells or rainwater for drinking and other domestic uses; these sources are extremely vulnerable to contamination and depletion following droughts and other extreme weather events, resulting in deteriorating community health. These impacts may be exacerbated by rising sea levels in coastal areas. Additional adverse health impacts which will particularly affect the most vulnerable population groups, such as older people, including increased frequency of heatwaves and incidence of vector-borne and other communicable diseases (Institute of Strategy and Policy on Natural Resources and Environment Viet Nam, 2009). To combat these risks, the Prime Minister approved a National Target Programme to respond to climate change in December 2008, in which the Ministry of Health was assigned to develop an action plan for the health sector in response to climate change between 2009 and 2015. The plan is based firmly on existing law and aims to respond to identified threats with understanding of the impacts of climate change on health. The specific objectives are to:

- evaluate disease models and the extent of climate change impacts on health;
- identify responses to climate change in the health sector;
- improve awareness among the community and health staff about health protection and climate change adaptation;
- consolidate mechanisms and policies, and strengthen organization;
- build capacity for health staff in response to climate change; and
- mainstream climate change response activities into the health sector's plan and activities.

The action plan includes over 35 distinct tasks and projects that respond to each of the above objectives that are to be implemented by 11 departments within the Ministry of Health. The total budget is approximately US$ 8 million (Ministry of Health, Viet Nam, 2010).

3.7.5 Vulnerability assessment

Important climate change and health impacts are associated with water-related natural hazards, particularly typhoons, floods and associated landslides, and droughts, as demonstrated by the country's climate change vulnerability map, highlighting the southern delta, north-west and coastal regions as particularly vulnerable (Yusuf & Francisco, 2010). Children, older people and other vulnerable groups are disproportionately affected both by these direct events and other indirect climate-associated impacts, such as those affecting food and water security. The poorest groups, which normally include ethnic minorities whose lifestyles are closely associated with seasonal patterns and whose adaptive capacity is relatively low, are also highly vulnerable. The traditional knowledge of such groups may, however, protect them to a degree from gradual climate change impacts and may hold significant value in planning climate change adaptation activities for others.

In common with other countries, agricultural and fishery sectors are highly vulnerable to climate change impacts, and disruptions to these sectoral activities will have health impacts. Storm surges and other causes of flooding may further disrupt the environmental determinants of health, exacerbating the transmission of waterborne diseases, reducing the availability of safe drinking-water and dispersing dangerous pollutants from industry. In Viet Nam, a high proportion of the population – including those residing in Ho Chi Minh City and those living close to sea level and in coastal areas – is especially vulnerable to flooding and associated health impacts (Bich et al., 2011; The World Bank Group, 2010; United Nations Environment Programme, 2009).

3.7.6 Governance and national activity on climate change and health

Viet Nam ratified the UNFCCC in 1994 and the Kyoto Protocol in 2002. The Ministry of Natural Resources and Environment (MONRE) is the national focal agency for activities related to climate change, and adaptation measures have been included in laws and plans, including the *National Strategy for Environmental Protection*, which includes measures for reducing the impact of sea level rise in coastal zones (Viet Nam, 2003). In early 2006, the MONRE-based International Support Group on Natural Resources and Environment (ISGE) established a climate change adaptation working group, which provides a forum for dialogue and promotes coordination for climate change adaptation measures (Fig. 31) (Chaudhry & Ruysschaert, 2007; Ministry of Natural Resources and Environment, Viet Nam, 2003).

Viet Nam has a long-standing institutional response system in place for natural disasters such as floods and typhoons as a consequence of the history of such events in the country. These disaster responses are key components of the country's climate change adaptation response. Specifically for early warning, real-time meteorological data are available and a 48-hour typhoon warning system is in place (Chaudhry & Ruysschaert, 2007). Viet Nam's policy framework for disaster management is set in the *Second National Strategy and Action Plan for Disaster Mitigation and Management 2001–2020* that prioritizes awareness-raising and participation, minimizing loss of life and assets. Additionally, the strategy includes establishment of disaster forecast centres in different areas of the country and specific coping mechanisms for floods, in addition to enhanced communications using advanced information and technology, and the use of schools and the media in awareness-raising (Ministry of Agriculture and Rural Development Viet Nam, 2001). These initiatives have obvious synergies with climate change and health adaptation. However, despite recent coverage of climate change challenges in the national media in recent years, accompanied by growing awareness of climate change impacts at the local level, a recent case study noted there is limited government ownership of an adaptive approach to future climate-related risks, and limited financing available for climate change adaptation (Chaudhry & Ruysschaert, 2007; United Nations Environment Programme, 2009).

Figure 31. Government institutional arrangements for responding to climate change

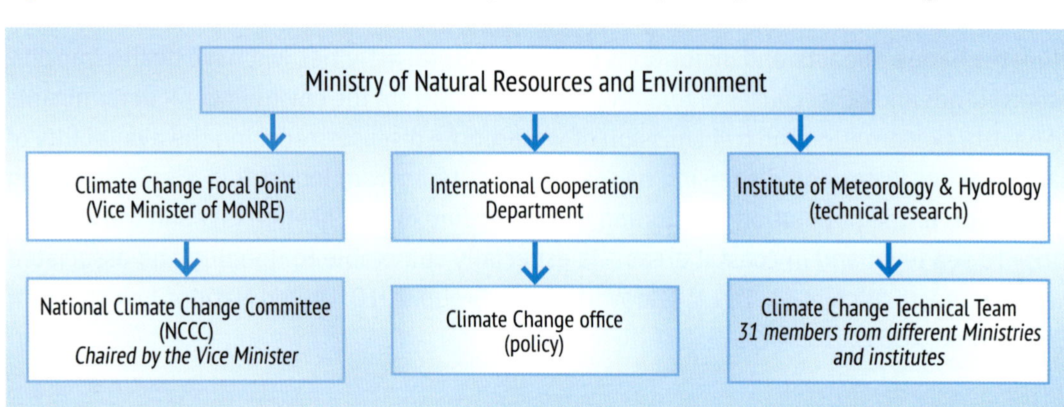

Source: Chaudhry & Ruysschaert, 2007.

3.7.7 Health adaptation activities

The Hanoi School of Public Health performed a quantitative study of the ability of hospitals in Bac Lieu, Phu Yen and Quang Ngai provinces to respond to climate change threats, such as natural disasters, increased incidence of infectious diseases and threats to the safety of other core systems. Recommendations included development of clear guidelines, strengthened human resource capacity and steps to ensure the continuation of essential supplies, such as blood and medical gases. The assessment tool could be further rolled out in different areas of the country to understand differing risk profiles, as necessary (Van & Trang, 2011). The Viet Nam Health Environment Management Agency (VIHEMA) under the Ministry of Health with technical and financial assistance from WHO has developed a *National Action Plan for Response to Health Impacts from Climate Change in Health Sector*. Guidelines on how to create action plans to respond to the health impacts of climate change at the provincial level were developed. Training on use of the guidelines was conducted in the two main regions of the country.

Additionally in 2010, with technical and financial assistance from WHO and the Viet Nam Health Environment Management Agency, the Research Center for Rural Population and Health developed a database on climate change and health and mapped out communities vulnerable to climate change impacts. This initiative will provide insights into the health impacts of climate change and help policy-makers reduce vulnerabilities (Research Center for Rural Population and Health Viet Nam, 2011).

North Gyeongsan Province, Republic of Korea

CHAPTER 4

Summary of work on climate change and health in the Western Pacific Region

4.1 Overview of climate change and health in the Western Pacific Region

The Western Pacific Region is home to approximately 1.8 billion people, roughly a quarter of the world's population and is exceptionally diverse in terms of geography, ethnicity and levels of economic development. It stretches over a vast area and includes highly developed countries such as Australia, Japan, New Zealand, the Republic of Korea and Singapore; and fast-growing economies such as China and Viet Nam. Many countries are highly exposed to climate hazards (Yusuf & Francisco, 2010). The range of climate change and health threats in the Region is also exceptionally diverse, reflecting geographical, epidemiological and developmental heterogeneity. In addition to the direct impacts of climate on health, when indirect pathways are considered almost all health outcomes are affected by climate change as a function of human behaviour, the climate vulnerability of public health facilities and impacts on the human food supply. Health practitioners working on climate change, therefore, attempt – without marginalizing other important topics – to use objective and accountable methods to define priority vulnerabilities within the health sector or which impact the health sector and for which adaptation can be planned and implemented.

4.2 Priority health risks of climate change

Despite geographical and climatic differences among Member States in the Western Pacific Region, priority thematic areas identified in climate change and health vulnerability assessments are similar, largely due to regional climate projections impacting related disease pathways. Similar population groups – those that are exposed, vulnerable and unable to adapt – are affected when environmental conditions interfere with their normal lifestyles. While this report describes threats and responses in only seven WHO Member States, similarities observed in this synthesis of country experiences may be considered demonstrative of challenges likely encountered in the Region as a whole.

The most direct impacts of weather on health outcomes are from extreme weather events such as storms, floods and landslides. Sometimes considered outside the scope of the health sector, an increased incidence of these events has the potential to cause health impacts in excess of those observed over recent decades. Projections of increased extreme weather events are clear and consistent. In Asia and the Pacific, many countries consist of archipelagos of hundreds or thousands of islands where climate change impacts are very keenly felt, as reflected in adaptation plans and frameworks. These impacts, however, are not confined to islands, and many countries are susceptible to extreme weather events that in recent years have placed pressure on services and have led to economic losses and displacement, migration and their associated health impacts.

The health impacts of extreme weather events in Member States are country- and region-specific. Coastal areas of Papua New Guinea, the Philippines and Viet Nam are particularly vulnerable to increasing incidence of cyclones and typhoons, rising sea levels and increased saltwater intrusion. These are perhaps the most obvious of climate change threats in the Western Pacific Region. However, extreme weather also brings health impacts in landlocked areas, including countries such as the Lao People's Democratic Republic and Mongolia that have no coast. Floods in low-lying areas of Cambodia and the Lao People's Democratic Republic have led to displacement of populations, reductions in food security and increased pressure on the health sector. These events also exacerbate existing issues of malnutrition, including lack of micronutrients. While rural populations are generally resilient to – and have experience in addressing – short-term weather changes and abnormalities, climate change refers to long-term changes that may, after several seasons of extreme conditions, lead to breakdowns in traditional systems, even in the most resilient population groups. This is exemplified by the *dzud*, a combination of drought and extreme cold, in Mongolia. Nomadic herders can survive losing half their herd during particularly cold winters but may be forced to seek urban employment if the event is repeated. Expanding populations live in camps made up of traditional Mongolian *gers*, or tents, in poverty and with inadequate infrastructure and service provision on the outskirts of the capital of Ulaanbaatar. These populations are vulnerable to health problems similar to those experienced by slum dwellers in other countries. Together with the coal-driven power plant and heavy traffic, the cooking and heating fires of those living in *gers* represent one of the major sources of air pollution in Ulaanbaatar.

For a number of reasons, the scientific literature on climate change and health indicates that vector-borne diseases are a priority health concern, and this is reflected in the adaptation plans and activities in the Region. All countries participating in this report mention vector-borne diseases in their vulnerability analyses, including Mongolia where they are generally not considered health priorities. Research in the area is ongoing and laboratory and modelling data support an increase in vulnerability to health outcomes if climate change affects the incidence, geographical range or seasonality of diseases. A significant impediment to making causative associations is the general lack of consistent and systematic epidemiological and vector surveillance data, particularly from the fringes of vector-borne disease transmission foci, where climate change-mediated expansions would be most visible. A number of initiatives are under way to improve vector-borne disease surveillance in the Region, including detecting drug-resistant pathogens and through laboratory strengthening. Such initiatives may provide data that will contribute to the understanding of climate change and other environmental impacts on disease epidemiology.

In recent years an increase in cross-border migration of pollutants, including Asian dust (a natural wind-carried event) in northeast Asia and Asian brown cloud or haze from South-east Asian forest fires affecting neighbouring countries, has been observed and the health impacts of these phenomena warrant additional scientific attention. Additional CO_2 emissions and human activities that may exacerbate or be indirectly caused by climate change may contribute to these and other environmental health risks in the Region.

4.3 Policy developments and progress

Those in the health sector are familiar with making decisions based on strong empirical evidence from epidemiological studies, clinical experience and expert recommendations. Climate science is a relatively new field whose most striking finding – that the climate is changing as a consequence of human activity and will likely continue – emerged only recently. Projecting the future climate has required increasingly complex computer models, and health impacts are inferred via extrapolations of health impacts of those projections. These risks may compete unfavourably with other more tangible health concerns, particularly in resource-constrained settings. However, the future health impacts of climate change are considerable, and these threats have been recognized by all Member States in World Health Assembly resolutions on climate change and health. National Adaptation Programmes of Action have been completed by the least developed countries (LDCs) and National Communications to the UNFCCC by low- and middle-income countries. All countries are beginning to develop National Adaptation Plans (NAPs).

Climate change and health is a relatively new area and the progress of most countries in terms of implementation of adaptation plans has been limited. However, progress has been made and themes are emerging from the Region in terms of vulnerability, adaptation and implementation of plans.

- There are advantages in utilizing a regional approach, with existing structures and networks, where countries face common burdens. This is particularly apparent where countries share geographical similarities or subregional associations (e.g. the Greater Mekong Subregion; Pacific island countries and areas).
- Climate change is being incorporated with other social determinants of health, such as migration and urbanization, which are rapidly altering the environmental determinants of health and are interacting with climate change-induced risks.
- Intersectoral institutional arrangements are being adopted to design and implement climate change and health plans, which may act as an entry point for other health topics requiring an intersectoral approach (neglected tropical diseases, emerging diseases, and water and sanitation).

Many countries have completed health sector vulnerability assessments and adaptation plans that go beyond the requirements of the UNFCCC and other agreements. Using intersectoral teams, countries have assessed which populations are most vulnerable to different kinds of health effects of climate change, identified weaknesses in the systems that should protect them and specified changes to respond to them. These assessments have also increased the profile of climate change and health and strengthened the case for investments in this area.

In the Western Pacific Region assessments were conducted by Member States, with support of WHO and comprised of teams of international experts assisting local ministries of health and environmental staff. Assessments focused on priority health areas in each country as identified during literature reviews, retrospective analyses, and site visits to assess adaptive capacity, exposure and susceptibility. Vulnerability according to different geographic areas and health risks was determined. Action plans were developed by national interdisciplinary working groups approved by ministries and national governments.

Specifically, WHO has developed a flexible process for vulnerability and adaptation assessments. The basic steps of an assessment are to: 1) frame and scope the assessment via determination of the regions and outcomes of interest and other parameters; 2) conduct a vulnerability assessment; 3) conduct an impact assessment of future risks and impacts; 4) develop an adaptation assessment to identify and prioritize policies and programmes to address current and projected health risks; and 5) establish an iterative process for monitoring and managing the health risks of climate change (WHO, 2013). This process is illustrated in Figure 32.

However, in addition to these multisectoral commitments and plans, in recent years the health sector in countries in the Western Pacific Region has become more active in health-specific climate change adaptation plans. These plans describe health priorities in Member States that are vulnerable to climate change and establish actions to be taken by policy-makers

Figure 32. Steps involved in a vulnerability and adaptation assessment

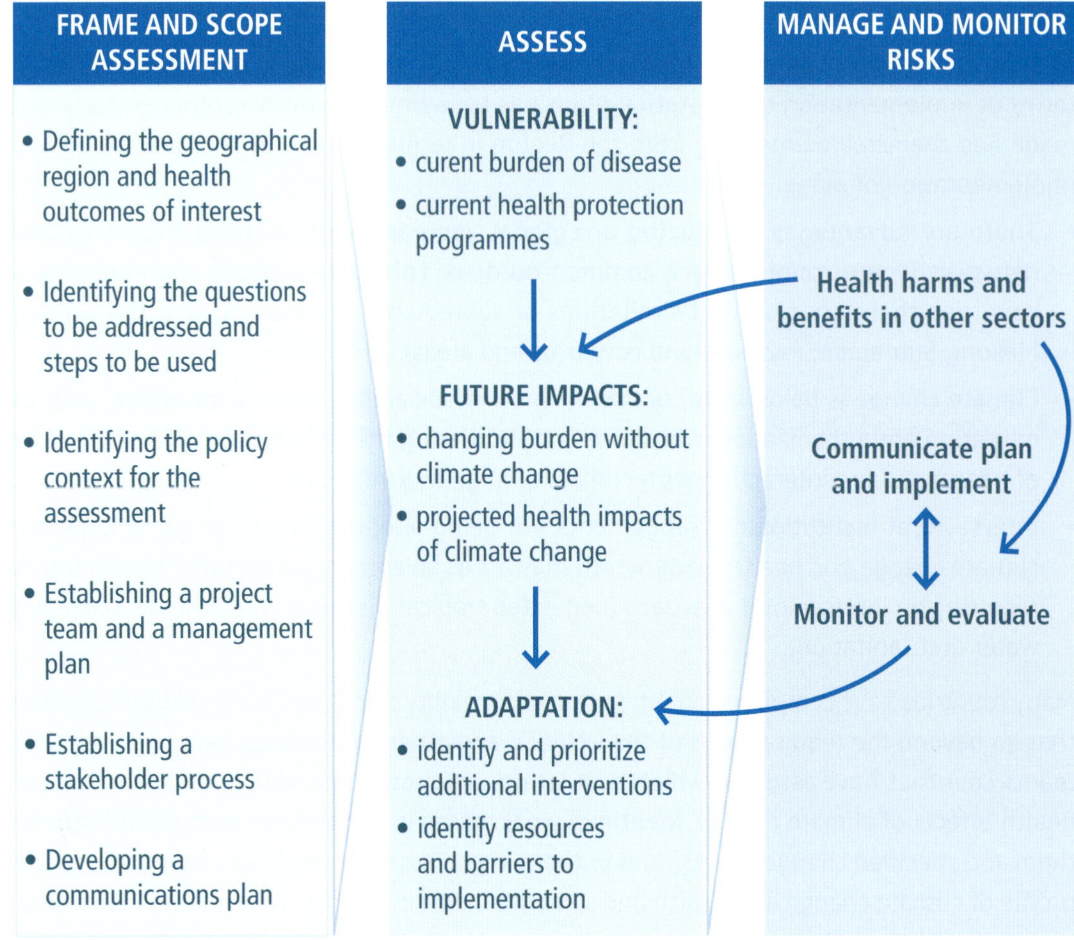

Source: modified from WHO, 2013.

based on objective evidence. Some of these country plans have been ratified as a whole by governments. Cambodia, for example, has recently developed a multisectoral *National Strategic Climate Change Plan*, containing a dedicated health component focusing on vector-borne and waterborne diseases, emergency preparedness and response, and improving the knowledge and research capacity on health impacts and vulnerability to climate change.

4.4 Implementation of adaptation plans

Widespread implementation of climate change and health adaptation plans has not yet taken place in any Member State of the Western Pacific Region, which is perhaps understandable as these plans were developed only recently and the activities they describe may compete with other health-care priorities. It should be noted that dedicated funds have been committed by developed countries to support adaptation in developing countries. While health sector utilization of these funds has been limited, they offer real potential for implemention of health adaptation plans.

In the Western Pacific Region, adaptation activities have taken place, managed by or with technical assistance from the WHO. Many of these funds originated from the Government of the Republic of Korea and have been partly supported by the Government of Japan, in line with the strategy of those governments to support "green growth" strategies in developing countries.

4.4.1 Strengthening control of vector-borne diseases to lessen the impact of climate change in the Western Pacific Region

WHO has recently supported a project on Strengthening Capacity to Respond to the Impacts of Climate Change on Vector-borne Disease Burdens both at the local and regional levels. The project was implemented in Cambodia, Mongolia and Papua New Guinea and funded through the East Asia Climate Partnership. It took place over 18 months, with six related outputs closely aligned to climate change and health adaptation plans. It offers a template for other similar initiatives aligned with national plans and strategies and utilizing climate change adaptation funding to strengthen vulnerable components of the health sector.

The outputs were:

- increased awareness and involvement of communities and stakeholders within and beyond the health sector in actions to minimize vector-borne disease consequences due to climate change;
- strengthened surveillance for vector-borne infections and climate change and capacity for rapid response to vector-borne disease outbreaks;
- strengthened capacity for vector control;
- strengthened capacity for effective diagnosis and treatment of vector-borne diseases;
- strategic information on knowledge gaps generated and utilized to better respond to climate change-induced vector-borne diseases; and
- strengthened country programmes and effective and efficient project management.

Vulnerability analyses were conducted in each country to identify areas, populations and vector-borne diseases at greatest risk of expansion and appropriate public health measures to combat them in each country. These activities were incorporated into each country's project work plan. Implementation began in 2011 and received strong political and institutional support in each country. As climate change is a highly cross-cutting issue, implementation at the country level was overseen by intersectoral technical working groups, including health, environmental, meteorological, agricultural, veterinary and other sectors. The project was managed by WHO in countries, and at the regional level experiences were exchanged among countries. Input from consultants from the Republic of Korea and other countries contributed to the information exchange and learning through training and interactions with local staff, tailored to individual national requirements. The project concluded:

- While it is difficult to determine what proportion of changes in the vector-borne disease burden is due to climate change, it is undeniable there will be changes that are likely to have negative impacts on the health of affected communities. Changes are likely to be heterogeneous.
- The project demonstrated a cross-programme and intersectoral approach that was successfully implemented on the ground in each of three countries vulnerable to the impacts of climate change.
- Political and sectoral enthusiasm was reflected in a willingness to implement the projects, and most project outputs were realized.
- The project approach on climate change and vector-borne diseases was original and comprehensive in terms of political commitment, institutional involvement, research elements, and the approach included a strong capacity-building element to address future threats. The project makes climate change interventions tangible, relevant and realistic.
- The project is a model for future development cooperation and a pathfinder for future climate change work, particularly in terms of adaptation.
- There were a number of limitations and challenges that are likely broadly reflective of climate change adaptation activities in other settings.
- The project duration was too short to demonstrate substantial changes at the outcome level. However, the planned outputs were realized. Due to the nature of climate change impacts, medium- and long-term approaches are required.
- The intersectoral project set-up and leadership have taken time to mature in some cases, and administrative agreements were time consuming.
- Due to the novel nature of climate change and health adaptation, there are some capacity limitations. These technical areas are new for governments, particularly ministries of health, in terms of research, awareness, and monitoring and evaluation of climate change interventions.

4.4.2 Project on the impact of climate change on water and health in vulnerable countries

The project aimed to strengthen country-level capacities in establishing better health surveillance of climate change and water and to support vulnerability assessments, water monitoring and assessments of the impact of climate change on water and health in Mongolia and Papua New Guinea. These countries represent extremes of the climate spectrum in the Western Pacific Region. The Ministry of Environment, Republic of Korea, contributed US$ 134 000 to the project. Main activities included:

- strengthening existing water quality monitoring systems in vulnerable areas;
- analysing the relationship between health outcomes and indicators of water availability (e.g. flow rate, water level, reserve capacity.) and quality (turbidity, hardness, salinity, NH_3, NO_2, NO_3, total bacteria, coliform, etc.);
- strengthening capacity-building in public health, water supply and climate sectors on the assessment and management of water monitoring and health issues;
- reviewing national adaptation programmes on water and health from the perspective of climate change impact; and
- proposing a surveillance system for monitoring the impact of climate change on water and health.

The quality of the municipal drinking-water supply did not meet acceptable standards in many areas, both urban and rural. Threats from microbial contamination of municipal water sources were evident in large cities in both Mongolia and Papua New Guinea. The high mineral content of water, including arsenic and fluoride, was prevalent in Gobi and other areas of rural Mongolia. Salinization of the drinking-water sources in relation to sea level rises and changes in the water table in the coastal areas of Papua New Guinea and in rural Mongolia were noted. Health status assessments suggested that drinking-water quality might be related to many health indicators, in addition to the incidence of gastrointestinal infectious diseases, both in rural and urban areas of Mongolia. Adaptation capacity to protect health from the water scarcity and water quality effects of climate change was not sufficient in either country, and the following recommendations were made:

- extend adaptation and mitigation activities on climate change and health to include water scarcity and quality issues;
- act legally and administratively to protect water sources and provide safe water to the public;
- conduct more extensive and comprehensive water quality monitoring of both municipal water supplies and water sources in smaller settlements;
- make water quality data publicly available and assess it regularly in relation to the health status of the people in the catchment area;
- strengthen and utilize health surveillance as a tool for monitoring the consequences of water quality, and develop a national information network able to integrate both health indicators and water quality and safety monitoring;
- strengthen the training of personnel both in the health and water sectors, and ensure adequate technical support to enable regular monitoring of the water quality;
- develop an "adequate technology" approach that is both sustainable and effective to ensure safe water sources and drinking-water quality is recommended; and
- strengthen the preparedness of the public health and administrative sector for water-related disasters and outbreaks of waterborne diseases.

The Red River, near Hanoi, Viet N

CHAPTER 5

Policy direction for the health sector's response to climate change

5.1 Introduction

This chapter provides a comprehensive outline of policy actions needed to address the health impacts of climate change and provides a structure for those actions. Tools that can be applied for the implementation of the policy actions are introduced. Priority actions for the health sector are listed, taking into account various levels of development.

5.2 Regional needs on climate change and health in the Western Pacific Region

In response to the rapid onset of climate change and its impact on health, the WHO Regional Committee for the Western Pacific, as noted in Chapter 1, endorsed the *Regional Framework for Action to Protect Human Health from Effects of Climate Change in the Asia Pacific Region* (WPR/RC59.R7) in 2008. The regional framework recommends action over three strategic objectives with the goal of building capacity and strengthening health systems in countries and of protecting human health from current and projected risks due to climate change at the regional level. The objectives are to:

- increase awareness of the health consequences of climate change;
- strengthen the capacity of health systems to provide protection from climate-related risks and substantially reduce the health sector's GHG emissions; and
- ensure that health concerns are addressed in decisions to reduce risks from climate change in other key sectors.

In support of the regional framework, WHO has supported Member States in their conduct of vulnerability and impact assessments and in developing national action plans and related strategies. Further regional work has been identified by Member States and experts as necessary, particularly to: i) develop and provide policy tools; ii) facilitate the development of evidence; iii) prioritize areas for action; iv) expand WHO technical support to all Member States; and v) establish a regional surveillance network for climate-sensitive diseases.

These needs are expanded upon below. Some of the items are described in detail in other sections of the document.

1. Develop and provide policy tools
(Refer to Section 5.5)

2. Facilitate the development of evidence

Despite strong theoretical linkages, attributing changing disease incidence to climatic determinants is a challenge. Developing this evidence should be a priority.

A number of activities should be encouraged to facilitate the process of developing evidence. These include:

- **The development and maintenance of longitudinal data collection systems** to enable monitoring of trends in disease incidence and other related determinants, such as vector distributions. The timescales of the current systems are insufficient to address decadal-scale events.
- **Developing regional capacity in statistical methods and environmental epidemiology** with a specific expertise in the analysis of climate and health data. A regional expert group would be able to assist with capacity development and guide implementation of projects, as well as provide credible data outputs and advocacy.[4]
- **Taking a regional approach to climate change research** will also enable broader understanding of climate change impacts and enhance credibility. The confidence of policy-makers – and the public – about health impacts will increase if the same effect is consistently observed among countries. The Association of Southeast Asian Nations (ASEAN) has attempted to nurture such an initiative under the Committee on Science and Technology, aiming to associate historical climatic factors with vector-borne disease incidence, incorporating other factors such as altitude using GIS.

3. Prioritize areas for action

Given finite budgets and competing priorities, it is important to prioritize actions based on climate change and health adaptation plans. In Cambodia, for example, the *Climate Change Strategic Plan for Public Health* (Ministry of Health Cambodia, 2012) builds on the national vulnerability analysis and adaptation plans, incorporating experience to date as well as national priorities.

The criteria selected by each country when refining implementation plans will likely vary, but an element of prioritization will be necessary to ensure plans remain focused, attractive to policy-makers and external donors, and timely.

4. Expand WHO technical support to Member States

It is important that the WHO Regional Office for the Western Pacific conduct a region-wide survey on climate change and health-related activities in all Member States. A full-range survey on the status and planning of national action plans for climate change, climate change and health, governance, awareness-raising, and strengthening health sector capacity is required. Mitigation activities, such as "greening" the health and environmental sectors and reducing GHG emissions, as well as the need for networking and capacity-building, should be included.

4. In the Western Pacific Region, an informal group of experts was recently convened and has been instrumental in the development of this report.

The survey will provide insights on the current status of Member States on climate change and health across the entire Region, providing an essential basis for priority-setting on regional action against climate change and health. It can also provide an effective means of communication and sharing of information and resources among Member States.

5. Establish a regional surveillance network for climate-sensitive diseases

Vector-borne disease control activities are mostly confined to specific countries and areas. Surveillance should be comprehensive and not confined to a specific disease or vectors, given that some vectors transmit multiple pathogens (e.g. *Aedes* mosquitoes). These approaches, however, are costly and require well-developed health-care infrastructure.

A useful approach to support comprehensive and robust surveillance and monitoring systems in the Region would be to encourage collaboration between countries to share data. This sharing of national level data can result in a clearer and stronger country and regional assessment of climate-sensitive diseases. A comprehensive surveillance network, incorporating an early warning system, can monitor the rate and range of outbreak across the subregion. A subregional network on vector-borne disease and climate change was recommended in 2011 by the WHO Regional Office for the Western Pacific.

For an effective and sustainable network, the following components are required:

- the active participation of neighbouring countries;
- adequate support provided for capacity-building, including data handling, management and network maintenance;
- development of appropriate protocols;
- local infrastructure to support the network's operation; and
- consistent funding.

For developed countries, mobilization of the network enabling an early warning system would improve predictability for tourism and industrial activities.

6. Seek regional adaptation funding

Countries of the Western Pacific Region are enthusiastic about implementation of adaptation plans that are seen as evidence-based, important and responsive to the most vulnerable populations. However, funding is often lacking or ad hoc. There is a clear need to develop proposals to attract substantial resources to fund implementation of adaptation plans. Such funding would include a number of co-benefits, increasing the visibility of health adaptation measures, enabling the mainstreaming of climate change into health activities, providing meaningful impacts, improving staff capacities and demonstrating the value of implementing climate change adaptation in the health sector, which has strong monitoring and evaluation capacity.

5.3 Strengthening and reforming health systems

The health sector's response to climate change impacts and health adaptation requires much more than simply introducing a climate–health programme in ministries of health. As discussed below, the "upstream" policy pathways exist to prevent the potential impacts of climate change on health, as well as within health systems. Health sector policy-makers may need to influence non-health sectors to maximize the health co-benefits of adaptation and mitigation policies in such areas as environment, transport, education, agriculture and energy, following health-in-all-policies and whole-of-government approaches (Bowen et al., 2013). Simultaneously, health sector policy-makers are advised to advocate for building stronger health systems, which will in turn increase the resilience of public health services to climate change.

5.3.1 Advocating for the health co-benefits of adaptation and mitigation measures

There are well-known health co-benefits arising from climate change mitigation and adaptation. A significant reason to mitigate climate-sensitive health risks is – in addition to health benefits from lowering air pollution – the potential to strengthen programmes addressing diarrhoeal infection, vector-borne disease, undernutrition, mental stress and cardiovascular diseases. Combined climate policies and health programmes in particular benefit the poorest and most vulnerable populations (WHO, 2014). For example, the potential health gains of a shift from private motorized transport to walking, cycling and rapid transit/public transport include reduced cardiovascular and respiratory disease from air pollution, less traffic injuries, and less noise-related stress (WHO, 2011a).

5.3.2 Strengthening pillars of the health system to increase resilience to climate change

It is vital that health sector adaptation to climate change is built on the identified pillars of health systems. This avoids duplication of efforts. The entry points for building a stronger health system resilient to climate change are:

- service delivery – effective, safe and good-quality health interventions provided in an efficient and equitable manner;
- health workforce – a high-performance health workforce is needed to achieve the best health outcomes possible;
- information – health information systems that ensure the production and application of reliable and timely information on health determinants, health systems performance and health status are essential for managing climate-related health risks;
- medical products and technologies – a range of medical products and technologies are needed to protect populations from climate-sensitive health conditions;
- financing – adequate funds are needed to maintain core health system functions, including in a crisis; and
- leadership and governance – political will to take action to address the health risks of climate change is essential (WHO, 2010c).[5]

5. Based on the Regional Office for Europe's Health Systems Crisis Preparedness Assessment Tool (WHO, 2010b)

5.3.3 Essential public health package for climate change resilience[6]

The following six public health services have been proposed as a minimum package for climate change resilience (WHO, 2010b). These services can be seen as complementary to those presented above, that strengthen health systems:

- comprehensive assessment of the risks posed by climate variability and change on population health and health systems;
- integrated environment and health surveillance;
- delivery of preventive and curative interventions for the effective management of identified climate-sensitive public health concerns;
- preparedness for, and response to, the public health consequences of extreme weather events, including population displacement;
- research; and
- strengthening of human and institutional capacities and intersectoral coordination.

Comprehensive assessments of the risks to population health and health systems

As presented earlier in this chapter, WHO has developed guidelines and tools for assessments of the risks posed by climate variability and change. These include tools for public health vulnerability and adaptation assessments and health systems assessments, as well as other tools for health risks, hazards and emergency capacity assessments.

Integrated environment and health surveillance

Timely decision-making and actions to predict and prevent the negative health effects of extreme weather events and environmental degradation, including those exacerbated by climate change, require further support in many countries. This challenge is impacted by a number of factors, including:

- fragmentation of surveillance activities;
- insufficient coordination among various established systems;
- low capacity to appropriately interpret integrated data; and
- lack of timely data for immediate decision-making.

Further to their risk and vulnerability assessments, as a second step in the resilience-building process, countries need functional and integrated environment and health surveillance systems. An essential function of such systems is to track environmental changes that affect health. These systems use a standardized set of environment and health indicators, including appropriate meteorological variables, and procedures to generate the required information for decision-makers and managers. An integrated environment and health surveillance system therefore builds on current integrated disease surveillance systems and expands them to incorporate key environmental indicators, including meteorological and climate data. Epidemic surveillance and preparedness for diseases that could emerge in new locations or populations due to climate and environmental changes are an integral part of this system.

6. This section is based on Annex 1 of WHO Consultation on the Essential Public Health Package to Enhance Climate Change Resilience (WHO, 2010c).

- **Delivery of preventive and curative interventions**

Vector-borne diseases, in particular malaria and dengue, as well as diarrhoeal and respiratory diseases, malnutrition and cardiovascular diseases are among the most climate-sensitive public health conditions identified by countries for immediate action. According to their respective local epidemiological circumstances and based on the conclusions of their vulnerability assessments, countries need to prioritize the public health programmes that require immediate strengthening to effectively limit potential increases in the incidence of climate-sensitive diseases. Such prioritization needs to be constantly reassessed based on evidence generated by the integrated environment and health information system. Countries will then be able to reduce the incidence of the disease conditions cited above by implementing initiatives, programmes and interventions such as:

- integrated vector management to reduce the incidence of malaria and other vector-borne diseases;
- water safety plans (in urban areas) and point-of-use water treatment (in slums and rural areas) to reduce the incidence of waterborne diseases;
- regulatory interventions to limit the concentrations of PM_{10} and $PM_{2.5}$ in ambient air to reduce the incidence of respiratory infections;
- food and nutrient supplementation; and
- other preventive interventions that are appropriate to local conditions for the most important climate-sensitive public health conditions.

- **Preparedness for and response to health consequences of emergencies and extreme events**

National and community health emergency management systems will have to be further developed in order to manage the health emergency risks associated with climate-related hazards, particularly in the context of climate change. Countries will need to review and where necessary update their capacities to ensure that the health sector can deal effectively with identified climate-sensitive hazards. Key actions will include:

- formulation and implementation of health emergency management policies;
- legislative frameworks and programmes;
- testing and updating of emergency response and recovery plans;
- deployment of early warning systems for health, including access to forecasts, as well as response and recovery operations, coordination and emergency communications;
- prevention and control of communicable diseases, mass casualty management, reproductive health, mental health and psychosocial support, environmental health, nutrition and emergency feeding, and fatality management;
- human resource development programmes for health emergency management, including training and education;
- community-based health risk-reduction programmes, including primary health care, first aid, health education and risk communications, early warning, and local emergency response planning;

- safer, resilient and prepared hospital programmes, including health facilities, critical infrastructure (such as water and sanitation), stockpiling of essential materials and ensuring sustainable health workforces in times of crisis (i.e. surge capacity); and
- integrated data management and surveillance systems.

- **Research**

There is insufficient understanding at the country level of the health effects of climate change on local populations. Each country will need to develop and implement a research agenda with two major objectives: (i) to comprehensively understand the local health effects of climate change; and (ii) to generate and disseminate knowledge of locally-appropriate adaptation measures while gaining momentum with respect to mitigation measures.

- **Strengthening core human and institutional capacities and intersectoral coordination**

Countries will be able to implement the above interventions in a reliable and effective manner only if the necessary core public health and environmental capacities are in place in terms of people and institutions. Gaps in these capacities at the country level must be identified and national capacity-building action plans prepared as part of future national adaptation programmes. Specific institutional coordination mechanisms will need to be established to ensure country ownership under the stewardship of ministries of health. Such mechanisms will be vital for planning, monitoring and evaluation of the national plans of action for the implementation of the proposed public health package. These mechanisms will also be responsible for ensuring intersectoral coordination and health representation in national and international development and in humanitarian and UNFCCC policy forums. Membership will be expanded beyond the relevant departments of the ministry of health to include representation from other sectors such as environment, agriculture, climate services, research and business.

5.4 Tools for evaluating climate change impacts on health and development of policy response

This report contains an overview of contemporary climate change and health initiatives planned and under way in the Western Pacific Region. Member States and environmental health stakeholders have made clear their commitment to address emerging threats. To help articulate these pathways and to provide practitioners with methods for communication and planning, we propose the use of two tools that are well-known in the field of environmental health: the Driving Force-Pressure-State-Exposure-Effect-Action (DPSEEA) framework and Health Impact Assessment (HIA). It is not proposed that these tools be a binding requirement. Rather, they are presented for discussion, modification and refinement, with the ultimate aim of providing users with a broad and useful range of methods and processes to facilitate the pursuit of climate change and health priorities in the Region. Also presented in this section is WHO's operational framework for building a climate-resilient health system (CRHS) that addresses concerns raised by Member States and partners on how the health sector and its operational basis – health systems – can effectively address the challenges increasingly presented by climate change.

5.4.1 DPSEEA framework

In order to prepare for and respond to the health risks and impacts of climate change effectively, it is important that causal pathways linking climate change and population health be appreciated and understood. The use of reliable indicators is also essential to assess and monitor the overall vulnerability and adaptation capacity of health systems to climate variability and change.

The DPSEEA framework is considered one of the most suitable ways to describe, design and assess activities for climate change and health (Hambling, Weinstein & Slaney, 2011). The DPSEEA framework describes environmental determinants of health from the highest upstream determinants to the eventual health outcomes, and it identifies key entry points for possible interventions at the societal level. The framework sees the most upstream determinants of health originating in driving forces (D), which lead to pressures on the environment (P). These pressures contribute to changes in the state (S) of the environment and human exposure (E) that have potential health effects (E). The framework identifies policy options and other actions (A), hence DPSEEA, which could be taken at each stage in the causal chain to alleviate the eventual adverse health effects. The framework is flexible and open to modification, as the national situations are different and the scientific knowledge is evolving constantly.

Figure 33. The DPSEEA framework for climate change and health

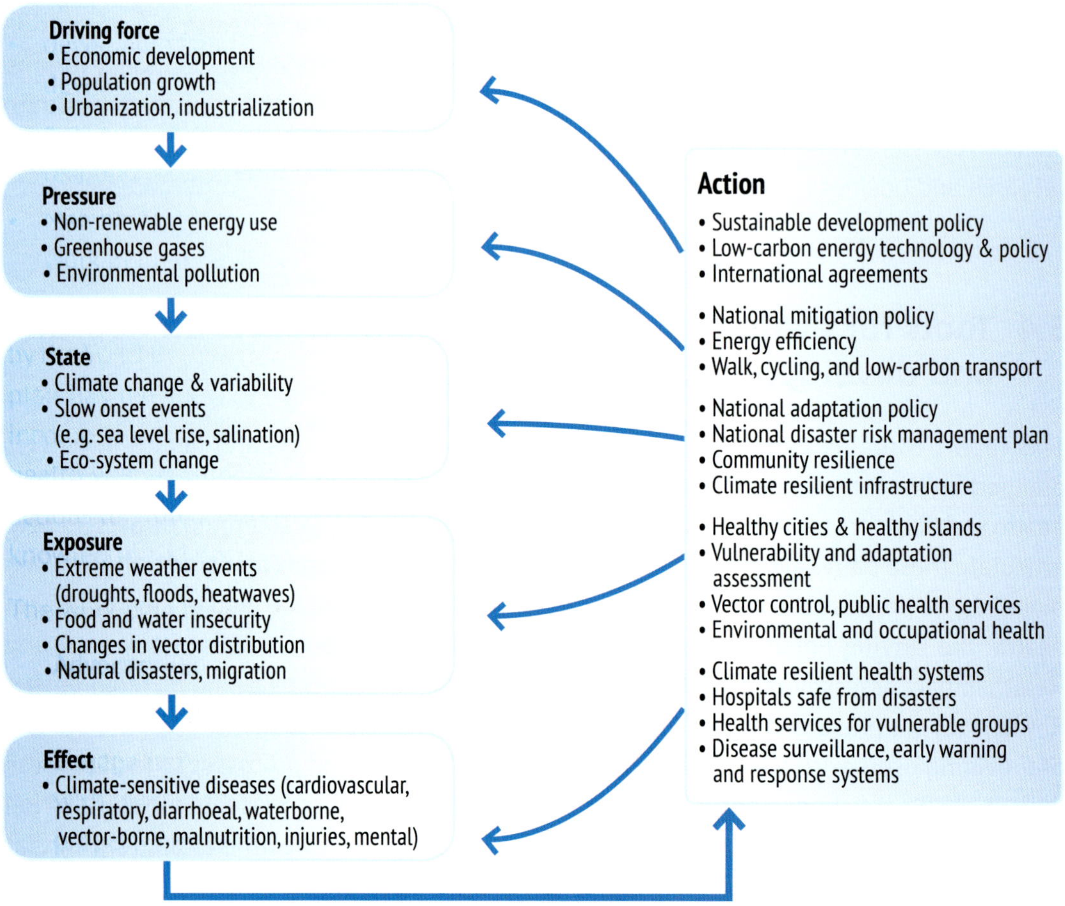

Sources: modified from Kovats et al., 2005; Kjellstrom & McMichael, 2013.

For Member States of the Western Pacific Region, a new version of the DPSEEA framework focusing on climate change and health is suggested as shown in Figure 33. This update reflects the situation and experiences in the Region (Kovats et al., 2005; Kjellstrom & McMichael, 2013).

The causal pathways of the DPSEEA framework provide policy-makers in the health sector with evidence-based policy options necessary not only to lead health sector adaptation at the downstream end of the cascade, but also to influence other sectors for primary prevention. It is important to use this framework in the context of health-in-all-policies, whole-of-government and whole-of-society approaches in order to maximize the health co-benefits of climate change adaptation.

Indicators for climate and health

Well-defined, measurable, reliable and relevant indicators are instrumental for policy-makers to assess and monitor human health vulnerability, aid in the design and targeting of interventions, and to measure the effectiveness of climate change adaptation activities. The WHO Regional Office for Europe has developed 17 indicators based on the DPSEEA framework to enable monitoring and assessment of environmental health issues related to climate change (WHO, 2010c). Table 7 shows those indicators and their position in the DPSEEA framework.

Table 7. The health-related indicators of global climate change developed and monitored by the WHO Regional Office for Europe

Topic areas		State	Exposure	Effect	Action
Extreme weather events	**Heatwaves**		Population exposure to heatwaves	Excess mortality due to heatwaves	Actions to prevent heat-related health effects
	Floods and droughts		Population exposure to actual floods Population vulnerability to floods		Actions to secure water supplies
Air quality	**Ambient air pollution**		Urban population exposure to ozone	Cardio-respiratory mortality	
	Airborne pollen allergens	Flowering of allergenic plants	Exposures to birch, alder and grass pollen Exposure to ragweed pollen	Antiallergy medication sales	
Infectious diseases	**Foodborne diseases**			Salmonellosis incidence and seasonality	Actions to prevent infectious diseases (cross-cutting)
	Waterborne diseases			Cryptosporidiosis incidence and seasonality	
	Vector-borne diseases		Lyme borreliosis occurrence of vector	Lyme borreliosis incidence	

Source: WHO, 2010c.

For countries in the Western Pacific Region, these indicators could be applied with minor adjustments as necessary. In the near future, if not developed already, countries of the Region may need their own set of indicators for climate change and its health impacts based on the DPSEEA framework in Figure 33 for monitoring of temporal and spatial trends and comparative analysis of national and regional situations.

5.4.2 Health impact assessment (HIA)

Context and rationale for use of HIA

The health sector has a direct incentive to ensure that measures taken on climate change adaptation (and mitigation) are those that make the greatest contribution to public health. A strategic harnessing of primary prevention opportunities offered by large-scale investments in climate change adaptation, particularly in policy domains that have the greatest potential to influence health determinants – such as urban planning, energy or agricultural policies – could generate substantial returns in terms of health. In addition, it is vital to include "outlying" factors that influence health, for example social and governance factors. Health is still not being systematically considered as part of decision-making in key policy domains that have the greatest potential to impact health outcomes and health determinants.

For countries to harness potential health co-benefits associated with climate change adaptation (and mitigation) in sectors other than the health sector, ministries of health must be able to assess, inform and influence decision-making in these other sectors. One of the key policy instruments available to support this is HIA.

Brief introduction to HIA

HIA is an analytical approach that identifies potential health and healthequity issues affected by a given policy or project, and provides possible options to mitigate, prevent or enhance those health outcomes (WHO European Centre for Health Policy, 1999; Quigley et al., 2006).

HIA is essentially comprised of six main steps (please refer to Brown, Pfeiffer & Lkhasuren, 2013 for more detail):

- **Screening**. A preliminary evaluation to determine whether a proposed policy, plan or project is likely to pose any significant health risks.
- **Scoping**. The process for outlining priority (expected) health concerns and the type of health assessment or HIA to be undertaken.
- **Analysis**. The systematic investigation, characterization and ranking/prioritization of the impacts that a policy, plan or project is likely to have on the health and well-being of communities.
- **Review (and communication) of the results of the HIA**. The results of the analysis and related recommendations are communicated back to the policy, plan or project proponent, key decision-makers, and potentially affected communities.
- **Development of the Public Health Action Plan (PHAP).** At this step, identified and prioritized impacts are translated into a public health action plan.

- **Implement and monitor the public health action plan.** This occurs during implementation of the proposed activities.

In the case of climate change policies, more systematic use of HIA to identify and harness health co-benefit opportunities may help to build wider and longer-lasting support for policy implementation.

Additional benefits of using HIAs in the context of policy-making for climate change, whether for adaptation or mitigation, may also include the following:

- HIA can be used to consider potential equity dimensions of proposed policies/measures.
- HIA can be used to establish an accountability framework for monitoring and measuring the health and/or social performance of climate change policies.
- HIA is both scientifically rigorous and participatory. Public participation and stakeholder engagement are essential.

Examples of HIA applied to climate change policies

WHO has been actively promoting the use of HIA in the context of climate change policy-making since the mid-2000s, particularly in the context of mitigation measures being promoted in economic sectors that have significant associated primary prevention opportunities such as energy, housing and transport.

From 2009 to 2011, WHO conducted an HIA of selected climate change mitigation measures proposed by the Intergovernmental Panel on Climate Change (IPCC) (IPCC, 2007a) in its Fourth Assessment Report. The analysis considered measures proposed in four sectors: residential buildings and housing; land transport; energy; and agriculture.[7] The findings, summarized in WHO's *Health in the Green Economy* series (WHO, 2013), point out several instances where measures focused solely on reducing GHG emissions either missed a major opportunity to address a public health issue (e.g. obesity) or aggravated an existing threat to health (e.g. air pollution). Two sector-specific examples include: i) the health benefits accrued from improving residential housing energy efficiency; and ii) in relation to land transport, where measures that were focused on reducing use of private motorized vehicles delivered health benefits.

Scaling up the use of HIAs in the context of climate change

Several initiatives are under way in the Western Pacific Region to support greater uptake and use of HIAs. One notable example is the intergovernmental Thematic Working Group on HIAs that was established in 2010 under the Regional Forum on Environmental Health in Southeast and East Asian Countries. HIA-related activities supported by the working group aim to share information, knowledge and tools on HIA practices and methods; support the establishment of HIAs as an integral part of decision-making processes in the Region; and facilitate capacity-building and exchanges among practitioners, such as through cooperative projects. In order for countries to maximize the sustained interest and financial resources

7. In each of the analyses, HIA steps used were limited to screening, scoping and analysis. The HIA process was applied to a global set of recommendations that was put forward by an international intergovernmental scientific committee. If adopted and implemented by a particular government (i.e. the respective measure would be applied to a particular national context and population), the additional steps in the HIA process, including the development, implementation and monitoring of the public health action plan, could be undertaken.

to support HIAs, existing HIA practices and related HIA capacity-strengthening efforts will need to address two key issues.

First, the focus of capacity development efforts in HIAs needs to move beyond a focus on developing HIA guidelines. While this is an integral part of influencing practice standards, experience from countries in the Western Pacific and South-East Asia regions that have been using HIAs for some time (e.g. Australia, New Zealand and Thailand) shows that the enabling environment and related support structures for HIAs are at least as important – if not more important – than the existence of regulations and best practice standards. For example, operational procedures may be needed to support the conduct of an HIA in a particular regulatory context, for instance if implemented as part of environmental impact assessment. There may be a need to establish a training facility to support the formation of locally qualified HIA practitioners. HIA regulators, i.e. those in government positions who are responsible for ensuring quality control over assessments conducted, may have a need for specific advice about how to relate the HIA process (and build capacity for its use) in the context of other sectoral decision-making processes, for example as part of spatial and/or urban planning.

Second, there needs to be greater coherence and alignment between HIA capacity-building efforts supported under the environmental health agenda and other HIA-related efforts supported under the global public health agenda, for example as part of wider initiatives in Health-in-all-Policies, social determinants of health, health equity and NCDs, particularly those focused on addressing NCD risk factors in urban settings. Articulating synergies between these different HIA-related efforts will not only ensure coherence in methods and processes being promoted, it may also be useful when building support and constituencies within the health sector, where unfortunately overall awareness and understanding about how to use HIAs as a policy-influencing instrument remains limited.

5.5 Actions for building climate-resilient health systems[8]

5.5.1 Scope and purpose

WHO's *Operational Framework for Building a Climate-Resilient Health Systems* (CRHS) (working draft) addresses concerns raised by Member States and partners on how the health sector and its operational basis (health systems) can effectively address the challenges increasingly presented by climate change. This framework has been conceived in light of regional climate projections, global lessons learnt in climate adaptation, and in support of the implementation of resolutions on climate and health approved by the WHO Regional Committee for the Western Pacific.

The objective of the framework is to guide health systems to become better prepared and capable of protecting health in an unstable and changing climate. Implementing the 10 key components[9] outlined in this framework will lead health organizations and authorities to consider how climate change affects their operations, and it could help them become better

8. This chapter is taken from Draft Summary of Operational Framework for Building Climate-Resilient Health Systems (working draft, Nov 7, 2013) prepared by Joy Guillemot (WHO Consultant), Diarmid Campbell-Lendrum and Elena Villalobos-Prats (WHO PHE).

9. Since its initial draft, the Operational Framework for Building Climate-Resilient Health Systems is now being reduced to six components: (1) governance and policy; (2) capacity development; (3) information and early warning systems; (4) service delivery; (5) essential products and technologies; and (6) financing.

able to anticipate, prevent, prepare for and manage climate-related health risks. These components align with the regional needs and areas to strengthen and reform the health system that were presented earlier in this chapter. This framework assists health managers in strengthening health operations in order to more effectively deliver and sustain health security in light of climate change, and fulfil national health commitments for climate action.

5.5.2 Overview of climate resilience

WHO has a working definition of a climate-resilient health system: a health system that can anticipate, respond to, cope with, recover from and adapt to climate-related shocks and stress so as to bring sustained improvements in population health, despite an unstable climate.

Climate resilience-building efforts take a systemic approach to complement health system strengthening by focusing on the system-wide capacities needed to address climate-specific health risks. Resilience is a useful approach because it is comprehensive and applicable to all climate-related health risks, and it supports an all-hazards approach to risk management by strengthening capacities that are useful for managing a range of health risks from disease outbreaks to health emergencies. It also enables multisectoral collaboration for prevention and management, and it empowers communities to play a role in assessments and responses.

The process of building resilience occurs in two principal ways: by reducing vulnerability, and by developing specific system capacities that improve the opportunities and choices available. This framework is intended to help develop the specific capacities that can enable a health system to become better prepared, responsive, adaptive, and more agile and efficient in light of climate change (Fig. 34).

Adaptation and climate resilience-building are closely related but not synonymous. Adaptation refers to strategies and measures put in place to respond to identified and predicted risks. Adaptation responses can help to build resilience. Resilience, broadly speaking, is about

Figure 34. Conceptual framework for resilience

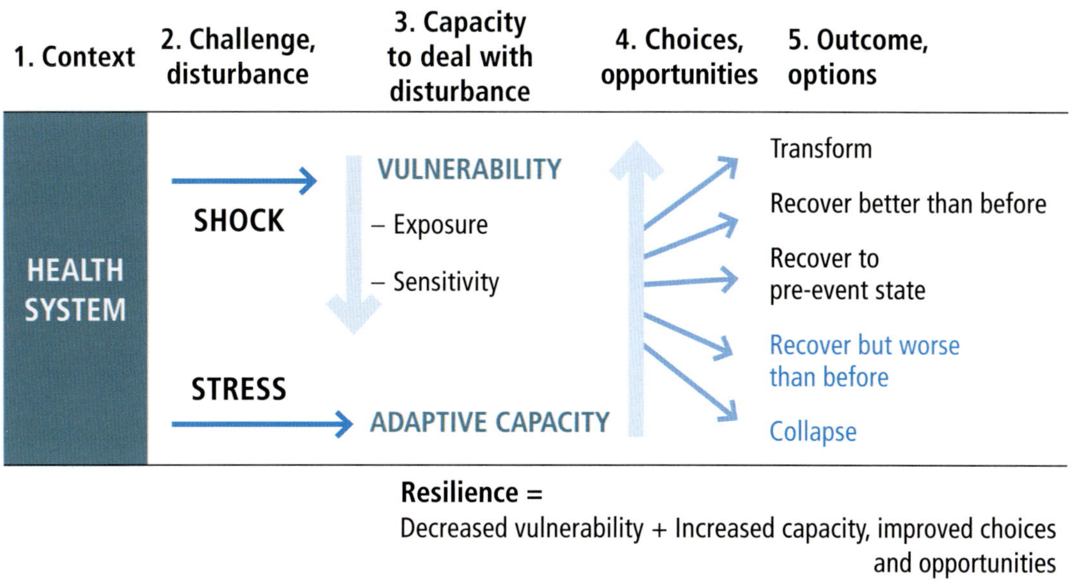

1. Context
2. Challenge, disturbance
3. Capacity to deal with disturbance
4. Choices, opportunities
5. Outcome, options

HEALTH SYSTEM
SHOCK
STRESS
VULNERABILITY
– Exposure
– Sensitivity
ADAPTIVE CAPACITY
Transform
Recover better than before
Recover to pre-event state
Recover but worse than before
Collapse

Resilience =
Decreased vulnerability + Increased capacity, improved choices and opportunities

Source: adapted by author from Defining Disaster Resilience: A DFID Approach Paper (DFID 2011).

strengthening the system as a whole to enable adaptation and abilities to manage change and stress effectively without catastrophic setbacks.

5.5.3 Overview of operational framework

The goal of the *Operational Framework for Building Climate-Resilient Health Systems* is to enhance the resilience of health systems. The objective is for health systems to become prepared and capable of protecting health in a changing climate.

Specifically, climate-resilient health systems are able to:

- recognize, monitor, anticipate, communicate and prepare for changing climate-related health risks, drawing upon and using the full spectrum of available knowledge and resources;
- prevent, respond to, manage and cope with uncertainty, adversity and stress;
- adjust and adapt operations in an innovative manner to changing risk conditions;
- recover from crises and setbacks with minimal outside support; and
- learn from experience and improve system capacity for the future.

Health policy and programming can be designed to build resilience to climate change. Ten key components have been identified as those that can facilitate the essential processes, information and partnerships to develop climate-resilient technical and institutional capacity within the health sector (Fig. 35). The proposed 10 components are drawn from existing WHO regional and national climate and health policies and programmes, and they identify steps essential to effectively address climate change. These strategic activities can help reduce health vulnerability, anticipate and monitor climate-specific health risks, build local capacity to manage change and crisis, improve flexibility and preparation to cope with stress and crisis, and manage change effectively.

In addition to the 10 components, it is crucial to emphasize that continued investment in primary health care and essential public health services is vital. In fact, it is the single most important investment that can be made to reduce population and health system vulnerability to climate change. In climate- vulnerable countries, it is paramount to continue improving the environmental and social determinants of health, including poverty and inequity reduction, as well as to strengthen core public health functions and health-care delivery systems.

5.5.4 The 10 components of the operational framework

Health systems which implement these 10 components can become progressively more climate-resilient by helping health actors attain the necessary competencies to identify and integrate climate-specific perspectives into their health operations. They also foster collaboration with communities and other sectors to improve prevention, preparedness and management of climate related risks.

- **Governance and policy**

Political will and good governance can pave the way for effective policies for health systems, climate and health. Management capacity to function under stress and changing conditions also is a critical part of climate resilience.

Figure 35. Operational framework for building climate-resilient health systems: main categories and their components

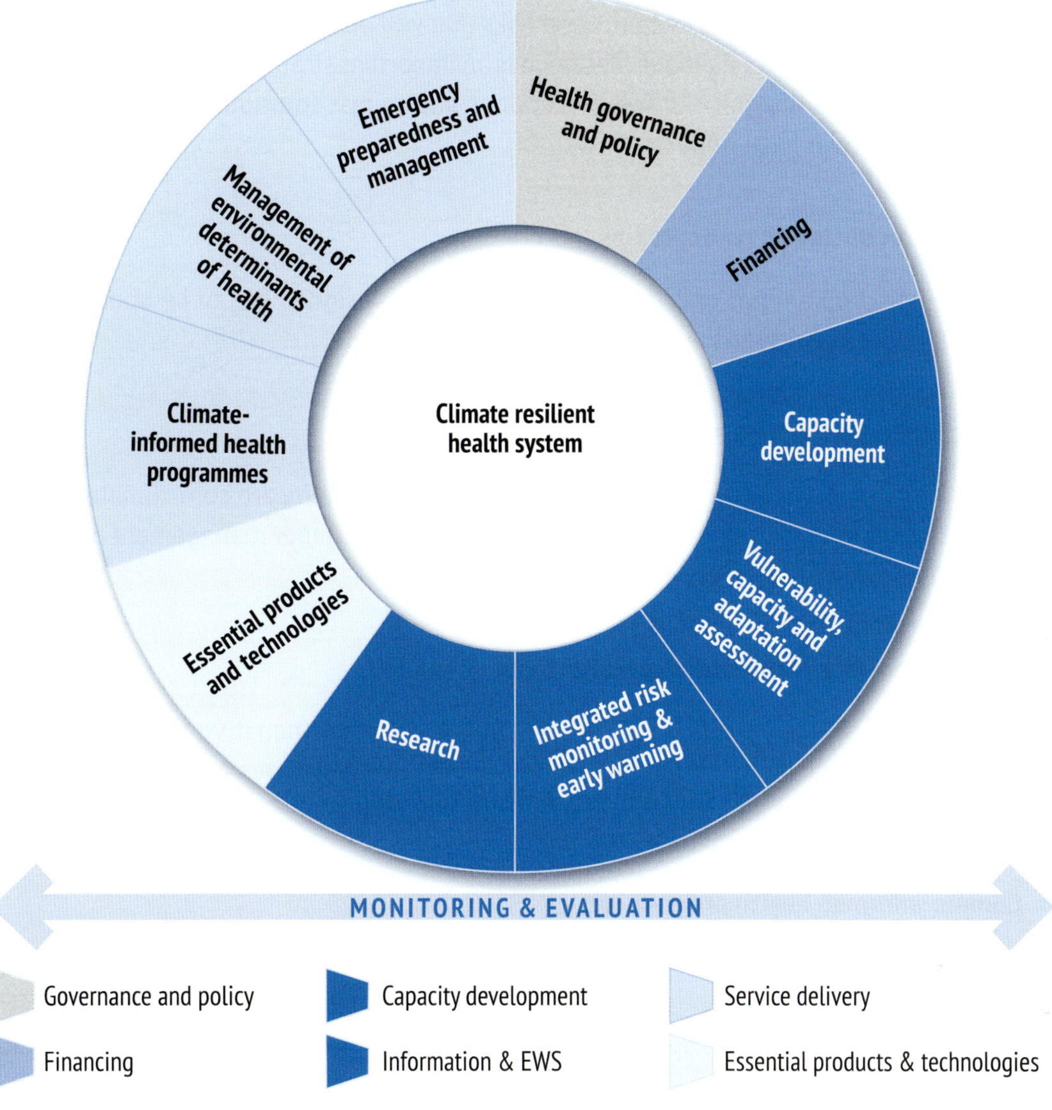

Source: WHO Operational Framework for Building Climate-resilient Health Systems: Working draft, Guillemot, et al.

- **Capacity development**

Institutional and human resource capacity development is a fundamental pathway for health systems to learn to become climate-resilient.

- **Vulnerability, capacity and adaptation assessment**

The comprehensive assessment of risks related to vulnerability, capacity and climate for both health systems and the population is a critical starting point for understanding the context and risks that require resilience.

- **Integrated risk monitoring and early warning**

Continuous monitoring of climate-related health risks and vulnerabilities, along with health service capacity, is essential to track trends in health outcomes and health service needs, identify changing conditions, and anticipate outbreaks and emergencies related to climatic conditions.

- **Research**

Local, national and international research can help clarify a range of unknowns about the local capacity, climate sensitivity, vulnerability and risks that will be unveiled through assessment, surveillance and community dialogue. Research is an important strategy to reduce uncertainty about changing local health risks and perceptions and to test solutions and opportunities to deal with potential local impacts.

- **Essential products and technologies**

The way health programmes operate and procedures are designed has to be updated so to ensure that those are responsive to the risks posed by climate.

- **Emergency preparedness and management**

Outbreaks and health emergencies triggered by climate variability are core climate-related health risks. Climate-smart emergency preparedness and emergency management are a vital part of building climate resilience.

- **Management of environmental determinants of health**

Scaling up multi-sectoral public health prevention can prevent negative health outcomes at their source and facilitate making earlier, more effective responses to deteriorating and changing environmental and climate-related risk conditions.

- **Climate-informed health programmes**

Health policy and programming must increasingly be designed and developed with climate risks and response capacity levels in mind. Climate information must be used to inform health decision-making and to apply adaptive management approaches to orient decisions according to future risk levels.

- **Financing**

Planning and management teams should consider that in addition to adequate development funds to maintain core health care and public health services, additional climate change specific funding mechanisms should be accessed so as to effectively build the resilience of the system.

CHAPTER 6

Conclusions

The human species has prospered over a hundred millennia through successful adaptation to the diversity of the global environment, including our changing climate. However, it is also evident that each adaptation process was accompanied by serious threats to the survival of certain tribes or species. The current threat of climate change began with the anthropogenic accumulation of GHGs, which poses a threat never before seen in human history. The health sector has a vital role to play in order to respond and minimize the threat that climate change poses to human health.

Scientific evidence has clearly disclosed the origins of climate change, clarified its pathway and predicted its future impact. Climate change is already apparent, and it presents us with a multitude of challenges and uncertainties. The health risks arising from climate change are many and varied, and their manifestation depends on factors including the local physical geography, socioeconomic status and population characteristics. Strengthening adaptive capacity and building the resilience of communities are key approaches to respond to the health risks posed by climate change. The capacity of the health sector to respond to changes in the climate and the health implications of these changes are among the most important factors in developing adaptive capacity.

While there are ample projections of future climate trends, quantitative predictions of the impact on health status due to climate change are not readily available. The consideration of probabilities, however, can present a reasonable approach to the problem. It is predicted that health impact in the near future will be enormous, even with an average temperature increase limited to 2 °C, which recent climate scenarios offer as a conservative estimate. There is serious concern that the impact of climate change could overwhelm the adaptive capacity of most countries.

The need for action is urgent. Action should be coordinated and networked across sectors and within and among countries in all regions. It should be effective and directed to the most vulnerable segments of the community. It must be comprehensive and mobilize the overall capacity of the community. Adaptation and mitigation efforts must occur in tandem, and their effectiveness will rely on the engagement of communities.

The countries of the Western Pacific Region have been proactive in response to the health impact of climate change. It is time to extend climate–health impact activities across all Member States and within and among diverse sectors. The commitment Member States have made to mitigate, prepare for and respond to the health impacts of climate change are commendable, and must be facilitated and supported by WHO and other partners. Fundamentally, human health must be mainstreamed in all efforts to address the impacts of climate change.

References

Ahlenius H, UNEP/GRID-Arendal. Historical trends in carbon dioxide concentrations and temperature, on a geological and recent time scale. Global Outlook for Ice and Snow, 2007 (http://www.grida.no/graphicslib/detail/historical-trends-in-carbon-dioxide-concentrations-and-temperature-on-a-geological-and-recent-time-scale_a210).

Allen BJ. Birthweight and environment in Tari. PNG Med J. 2002;45(1-2):88-98.

Arrhenius S. On the influence of carbonic acid in the air upon the temperature of the ground. Philos Mag. 1896;5(41): 237–76.

Asia Pacific Observatory on Health Systems and Policies, World Health Organization Regional Observatory on Health Systems and Policies. The Philippines Health System Review. Health Systems in Transition. Geneva: World Health Organization; 2011;2(1).

Asian Development Bank. Papua New Guinea's Strategic Program for Climate Resilience. 2012;1-39 (http://www.climateinvestmentfunds.org/cif/sites/climateinvestmentfunds.org/files/Strategic_Program_for_Climate_Resilience_for_Papua_New_Guinea.pdf, accessed 11 Oct 2013).

Bai L, Morton LC, Liu Q. Climate change and mosquito-borne diseases in China: a review. Global Health. 2013;9:10. doi: 10.1186/1744-8603-9-10.

Bambrick HJ, Capon AG, Barnett GB, Beaty RM, Burton AJ. Climate change and health in the urban environment: adaptation opportunities in Australian cities. Asia Pac J Public Health. 2011;23(2 Suppl):67S–79.

Banu S, Hu W, Hurst C, Tong S. Dengue transmission in the Asia-Pacific region: impact of climate change and socio-environmental factors. Trop Med Int Health. 2011;16(5):598-607. doi: 10.1111/j.1365-3156.2011.02734.x.

Beebe NW, Cooper RD, Mottram P, Sweeney AW. Australia's dengue risk driven by human adaptation to climate change. PLoS Negl Trop Dis. 2009;3(5):e429.

Berlioz-Arthaud A, Guillard B, Goarant C, Hem S. Hospital-based active surveillance of human leptospirosis in Cambodia. Bull Soc Pathol Exot. 2010;103(2):111-8. doi: 10.1007/s13149-010-0043-2. [French].

Berry HL, Bowen K, Kjellstrom T. Climate change and mental health: a causal pathways framework. International Journal of Public Health. 2010;55(2):123-132. doi: 10.1007/s00038-009-0112-0.

Bi P, Williams S, Loughnan M, Lloyd G, Hansen A, Kjellstrom T, Dear K, Saniotis A. The effects of extreme heat on human mortality and morbidity in Australia: implications for public health. Asia Pac J Public Health. 2011;23(2 Suppl):27S-36. doi: 10.1177/1010539510391644.

Bich TH, Quang LN, Ha le TT, Hanh TT, Guha-Sapir D. Impacts of flood on health: epidemiologic evidence from Hanoi, Vietnam. Glob Health Action. 2011;4:6356. doi: 10.3402/gha.v4i0.6356. Epub 2011 Aug 23.

Birley M. Health impact assessment, integration and critical appraisal. Impact Assessment and Project Appraisal. 2003;21(4):313-321. doi: 10.3152/147154603781766158.

Bowen KJ, Ebi K, Friel S, McMichael AJ. A multi-layered governance framework for incorporating social science insights into adapting to the health impacts of climate change. Glob Health Action. 2013; Sep 10;6:21820. doi: 10.3402/gha.v6i0.21820.

Brown O, Pfeiffer M, Lkhasuren O. Good practice guidance and minimum quality standards for health impact assessment in Mongolia. Ulaanbaatar: Ministry of Health, Mongolia; 2013

Burkart K, Khan MH, Krämer A, Breitner S, Schneider A, Endlicher WR. Seasonal variations of all-cause and cause-specific mortality by age, gender, and socioeconomic condition in urban and rural areas of Bangladesh. Int J Equity Health. 2011;10(1):32.

Buschmann J1, Berg M, Stengel C, Sampson ML. Arsenic and manganese contamination of drinking water resources in Cambodia: coincidence of risk areas with low relief topography. Environ Sci Technol. 2007;41(7):2146-52.

Cann KF, Thomas DR, Salmon RL, Wyn-Jones AP, Kay D. Extreme water-related weather events and waterborne disease. Epidemiol Infect. 2013;141(4):671-86. doi: 10.1017/S0950268812001653.

Centers for Disease Control and Prevention. Nutritional assessment of children after severe winter weather – Mongolia, June 2001. MMWR. 2002;51(01):5-7.

Centers for Disease Control and Prevention, Republic of Korea. Communicable Diseases Response Center. [Water & foodborne disease outbreaks in Korea, 2007-2009]. Osong: Centers for Disease Control and Prevention, Republic of Korea; 2010;3(26):2. [Korean] (Translated by editors). Centers for Disease Control and Prevention, Republic of Korea. Climate Change and Human Health: Impact and Adaptation Strategies in the Republic of Korea. 2010.

Centers for Disease Control and Prevention, Republic of Korea. Trends in water- and foodborne outbreaks in Korea, 2007–2009. Public Health Weekly Report. 2010;26(3):429-431.

Chan M. Cutting carbon, improving health. Lancet. 2009;374(9705):1870-1. doi: 10.1016/S0140-6736(09)61993-0.

Chaudhry P, Ruysschaert G. Climate change and human development in Viet Nam. Human Development Report. 2007;46:1–18.

Cheng JJ, Berry P. Health co-benefits and risks of public health adaptation strategies to climate change: a review of current literature. Int J Public Health. 2013;58(2):305-11. doi: 10.1007/s00038-012-0422-5.

Chung JY, Honda Y, Hong YC, Pan XC, Guo YL, Kim H. Ambient temperature and mortality: an international study in four capital cities of East Asia. Sci Total Environ. 2009;408(2):390–6. doi: 10.1016/j.scitotenv.2009.09.009.

Climate Change Commission. National Framework Strategy on Climate Change 2010–2022. Manila: Office of the President of the Philippines; 2010.

Climate Change Commission. National Climate Change Action Plan executive summary. Manila: Office of the President of the Philippines; 2011.

Commonwealth Health Ministers' Update. Country Survey on Health and Climate Change. Papua New Guinea. 2009: 50-3.

CRED [website]. Centre for Research on the Epidemiology of Disasters; 2013 (http://www.cred.be/, accessed 7 June 2013).

Cruz RV, Harasawa H, Lal M, Wu S, Anokhin Y, Punsalmaa B et al. Asia. In: Parry ML, Canziani OV, Palutikof JP, van der Linden PJ, and Hanson CE, editors. Climate Change 2007: impacts, adaptation and vulnerability. Contribution of Working Group II to the Fourth Assessment Report of the Intergovernmental Panel on Climate Change. Cambridge: Cambridge University Press; 2007:469–506.

Danish International Development Agency (DANIDA), Ministry of Foreign Affairs of Denmark. Climate change initiatives. 2013 (http://vietnam.um.dk/en/danida-en/climate-change-initiatives/, accessed 11 January 2015).

Department of Health, Philippines and Office of the World Health Organization Representative in the Philippines. Implementation of the MDGF 1656 Health Component. Joint Program on Strengthening the Philippines Institutional Capacity to Adapt to Climate Change in Health (2009–2011). Manila: WHO Philippines; 2011 (http://www.wpro.who.int/philippines/areas/emergencies_disasters/speed/en, accessed 13 Oct 2013).

Department of Health, Philippines and World Health Organization. Climate change and health news. 2012(1) (http://www.wpro.who.int/philippines/publications/climate_change_newsletter_final.pdf, accessed 11 January 2015).

Department of Health, Philippines and World Health Organization. Policies handbook on climate change and health. 2012.

DFID. Defining disaster resilience: a DFID approach paper. London: Department for International Development; 2011.

Desvars A, Cardinale E, Michault A. Animal leptospirosis in small tropical areas. Epidemiol Infect. 2011;44:167–188. doi: 10.1017/S0950268810002074.

Feldman PR, Rosenboom JW, Saray M, Navuth P, Samnang C, Iddings S. Assessment of the chemical quality of drinking water in Cambodia. J Water Health. 2007;5(1):101-16.

Few R, Ahem M, Matthies F, Kovats S. Floods, health and climate change: a strategic review. Working Paper No. 63. Tyndall Centre for Climate Change Research. 2004.

Food and Agriculture Organization. Assessment of the contribution of forestry to poverty alleviation in Lao People's Democratic Republic (http://www.fao.org/docrep/016/i2732e/i2732e02.pdf, accessed 11 June 2013).

Githeko AK, Lindsay SW, Confalonieri UE and Patz JA. Climate change and vector-borne diseases: a regional analysis. Bull World Health Organ. 2000;78(9):1136–47. (http://www.pubmedcentral.nih.gov/articlerender.fcgi?artid=2560843&tool=pmcentrez&rendertype=abstract).

Gomboluudev P. Country Report of Mongolia. 2007.

Government of the Lao People's Democratic Republic. The First National Communication on Climate Change. Vientiane; 2000 (http://unfccc.int/resource/docs/natc/laonc1.pdf, accessed 29 Sep 2009).

Government of the Lao People's Democratic Republic. The Ketsana typhoon in the Lao People's Democratic Republic – Damage, loss and needs assessment. 2009.

Government of Papua New Guinea. Papua New Guinea National Health Plan 2011–2020. 2010;1.

Government of the Philippines. The Philippines' Initial National Communication on Climate Change. 1999.

Guillemot J, Campbell-Lendrum D, Villalobos-Prats E. Draft summary of operational framework for building climate resilient health system (working draft, Nov 7, 2013).

Ha J, Kim H. Changes in the association between summer temperature and mortality in Seoul, South Korea. Int J Biometeorol. 2013;57(4):535-44. doi: 10.1007/s00484-012-0580-4.

Haines A, Kovats RS, Campbell-Lendrum D, Corvalan C. Climate change and human health: impacts, vulnerability, and mitigation. Lancet. 2006;367(9528):2101-9.

Hajat S, Kosatky T. Heat-related mortality: a review and exploration of heterogeneity. J Epidemiol Community Health. 2010;64(9):753–60. doi: 10.1136/jech.2009.087999.

Hambling T, Weinstein P, Slaney D. A review of frameworks for developing environmental health indicators for climate change and health. Int J Environ Res Public Health. 2011;8(7):2854-75. doi: 10.3390/ijerph8072854.

Handmer J, Honda Y, Kundzewicz Z, Arnell N, Benito G, Hatfield J et al. Changes in impacts of climate extremes: human systems and ecosystems. In: Field C, Barros V, Stocker T, Qin D, Dokken D, Ebi KL et al, editors. Managing the risks of extreme events and disasters to advance climate change adaptation: Intergovernmental Panel on Climate Change (IPCC) Special Report. Cambridge and New York: Cambridge University Press; 2012:231–90.

Hansen J, Ruedy R, Sato M, Lo K. Global surface temperature change. Rev. Geophysics. 2010;48(4): RG4004. doi: 10.1029/2010RG000345.

Harris PJ, Harris E, Thompson S, Harris-Roxas B, Kemp L. Health and wellbeing in environmental impact assessment in New South Wales, Australia: Auditing health impacts within environmental assessments of major projects. Environmental Impact Assessment Review. 2009;29(5):310-8.

Hashizume M, Armstrong B, Hajat S, Wagatsuma Y, Faruque ASG, Hayashi T, Sack DA. The effect of rainfall on the incidence of cholera in Bangladesh. Epidemiology. 2008;19:103–110.

Hashizume M, Terao T, Minakawa N. The Indian Ocean Dipole and malaria risk in the highlands of western Kenya. Proc Natl Acad Sci U S A. 2009;106(6):1857-62. doi: 10.1073/pnas.0806544106.

Hashizume M, Faruque AS, Terao T, Yunus M, Streatfield K, Yamamoto T, et al. The Indian Ocean dipole and cholera incidence in Bangladesh: a time-series analysis. Environ Health Perspect 2011;119(2):239-44.

Honda Y, Ono M. Issues in health risk assessment of current and future heat extremes. Glob Health Action. 2009;2. doi: 10.3402/gha.v2i0.2043.

Hosking J, Mudu P, Dora C. Health co-benefits of climate change mitigation – transport sector. Geneva: World Health Organization; 2011.

Inglis RF, Gardner A, Cornelis P, Buckling A. Spite and virulence in the bacterium Pseudomonas aeruginosa. Proc Natl Acad Sci U S A. 2009;106(14):5703-7. doi: 10.1073/pnas.0810850106.

Imai C, Hashizume M, Cheong HK, Kim H, Honda Y, Eum JH, Kim CT, Kim J, Kim Y, Fengthong T. The impacts of global and local climates on dengue fever in Lao PDR and Cambodia. The 25th International Society for Environmental Epidemiology, 2013.

Institute of Strategy and Policy on Natural Resources and Environment Viet Nam. Viet Nam Assessment Report on Climate Change. Ha Noi, 2009.

Intergovernmental Panel on Climate Change. Climate change: the IPCC 1990 and 1992 Assessments_overview; 1992 (http://www.ipcc.ch/ipccreports/1992%20IPCC%20Supplement/IPCC_1990_and_1992_Assessments/English/ipcc_90_92_assessments_far_overview.pdf).

Intergovernmental Panel on Climate Change. Climate Change 2013: the physical science basis. Contribution of Working Group I to the Fifth Assessment Report of the Intergovernmental Panel on Climate Change [Stocker TF, Qin D et al., editors]. Cambridge: Cambridge University Press; 2013.

Intergovernmental Panel on Climate Change. Climate Change 2014: impacts, adaptation, and vulnerability part A: global and sectoral aspects. Contribution of Working Group II to the Fifth Assessment Report of the Intergovernmental Panel on Climate Change [Field CB, Barros VR et al., editors]. Cambridge: Cambridge University Press; 2014.Intergovernmental Panel on Climate Change. Fourth Assessment Report. Cambridge and New York: Cambridge University Press; 2007a.

Intergovernmental Panel on Climate Change. Fourth Assessment Report. Working Group I: The Physical Science Basis; 2007b (https://www.ipcc.unibe.ch/publications/wg1-ar4/faq/wg1_faq-1.3.html).

Intergovernmental Panel on Climate Change. Intergovernmental Panel on Climate Change summary for policymakers: emissions scenarios. 2000.

Intergovernmental Panel on Climate Change. Managing the risks of extreme events and disasters to advance climate change adaptation. Special report of the Intergovernmental Panel on Climate Change [Field CB, Barros V, Stocker TF, Dahe Q et al., editors]. Cambridge: Cambridge University Press; 2012.

International Energy Agency. CO_2 emissions from fuel combustion (2012 Edition). Paris (http://www.iea.org/termsandconditionsuseandcopyright/, accessed 12 December 2013).

Ivanova S, Herbreteau V, Blasdell K, Chaval Y, Buchy P, Guillard B, Morand S. Leptospira and rodents in Cambodia: environmental determinants of infection. Am J Trop Med Hyg. 2012;86(6):1032-8. doi: 10.4269/ajtmh.2012.11-0349.

Jaenson TG, Lindgren E. The range of Ixodesricinus and the risk of contracting Lyme borreliosis will increase northwards when the vegetation period becomes longer. Ticks Tick Borne Dis. 2011;2(1):44-9.

Joint Climate Change Initiative. Cambodia and climate change: a brief review of climate change responses in Cambodia. 2010;1–23 (http://ppcrcambodia.files.wordpress.com/2012/11/solar-1002-cc-review-cam.pdf).

Jutla AS, Akanda AS, Griffiths JK, Colwell R, Islam S. Warming oceans, phytoplankton, and river discharge: implications for cholera outbreaks. Am J Trop Med Hyg. 2011;85(2)303-308. doi: 10.4269/ajtmh.2011.11-0181 (http://drinkingwateradvisor.com/2011/08/10/jutla-et-al-2011-warming-oceans-phytoplankton-and-river-discharge-implications-for-cholera-outbreaks/).

Kasper MR, Blair PJ, Touch S, Sokhal B, Yasuda CY, Williams M et al . Infectious etiologies of acute febrile illness among patients seeking health care in south-central Cambodia. Am J Trop Med Hyg. 2012;86(2):246-53. doi: 10.4269/ajtmh.2012.11-0409.

Kasper MR, Sokhal B, Blair PJ, Wierzba TF, Putnam SD. Emergence of multidrug-resistant Salmonella enterica serovar Typhi with reduced susceptibility to fluoroquinolones in Cambodia. Diagn Microbiol Infect Dis. 2010;66(2):207-9. doi: 10.1016/j.diagmicrobio.2009.09.002.

Kim SH. Relationship between hospital visits of adult allergy patients sensitized by tree pollens and climatic factors: Presentation materials of outcome and issues of research on health effects of climate change at the comprehensive academic forum of the Forum for Climate Change & Health in Korea, 2010.

Kim YM, Kim S, Cheong HK, Ahn B, Choi K. Effects of heat wave on body temperature and blood pressure in the poor and elderly. Environ Health Toxicol. 2012b;27:e2012013 (http://dx.doi.org/10.5620/eht.2011.26.e2011009).

Kim YM, Park JW, Cheong HK. Estimated effect of climatic variables on the transmission of Plasmodium vivax malaria in the Republic of Korea. Environ Health Perspect. 2012a;120:1314-9.

Kjellstrom T, McMichael AJ. Climate change threats to population health and well-being: the imperative of protective solutions that will last. Glob Health Action. 2013;6:20816. doi: 10.3402/gha.v6i0.20816.

Kolstad EW, Johansson KA. Uncertainties associated with quantifying climate change impacts on human health: a case study for diarrhea. Environ Health Perspect. 2011;119(3):299-305.

Kondo M, Honda Y OM. Growing concern about heatstroke this summer in Japan after Fukushima nuclear disaster. Environ Health Prev Med. 2011;16(5):279–80. doi: 10.1007/s12199-011-0228-8.

Kovats RS, Campbell-Lendrum D, Matthies F. Climate change and human health: estimating avoidable deaths and disease. Risk Anal. 2005;25(6):1409-18.

Kovats RS, Campbell-Lendrum D, McMichael AJ, Woodward A, Cox JS. Early effects of climate change: do they include changes in vector-borne disease? Philos Trans R Soc Lond B Biol Sci. 2001;356(1411):1057-68.

Kovats RS, Hajat S. Heat stress and public health: a critical review. Annu Rev Public Health. 2008;29:41–55.

Lao People's Democratic Republic. National Adaptation Programme of Action to Climate Change. Development. 2009.

Laras K, Cao BV, Bounlu K, Nguyen TK, Olson JG, Thongchanh S et al. The importance of leptospirosis in Southeast Asia. Am J Trop Med Hyg. 2002;67(3):278-86.

Lau CL, Skelly C, Smythe LD, Craig SB, Weinstein P. Emergence of new leptospiral serovars in American Samoa - ascertainment or ecological change? BMC Infect Dis. 2012;12:19. doi: 10.1186/1471-2334-12-19.

Mandakh N. Desertification map. Tsogtbaatar J, Khudulmur S, editors. Desertification atlas of Mongolia. Ulaanbaatar: Admon; 2012. p.89-95.

Mandakh N, Dash D, Khaulenbek A. Geo-ecological issues in Mongolia. In: Tsogtbaatar J, editor. Ulaanbaatar. 2007;63–73.

Manga L, Bagayoko M, Meredith T, Neira M. Overview of health considerations within National Adaptation Programmes of Action for climate change in least developed countries and small island states. Geneva: World Health Organization; 2010.

Manila Observatory. Technical Primer on Climate Change in the Philippines [Table of Contents]. March 2010.

Martens WJ, Niessen LW, Rotmans J, Jetten TH, McMichael AJ. Potential impact of global climate change on malaria risk. Environmental Health Perspectives. 1995;103(5):458–64 (http://www.pubmedcentral.nih.gov/articlerender.fcgi?artid=1523278&tool=pmcentrez&rendertype=abstract).

McMichael AJ, Friel S, Nyong A, Corvalan C. Global environmental change and health: impacts, inequalities, and the health sector. BMJ. 2008;336(7637):191-4.

McMichael AJ, Woodruff RE, Hales S. Climate change and human health: present and future risks. Lancet. 2006;367(9513):859-69.

Mekong River Commission. Climate Change Adaptation Initiative 2011–2015 Programme Document. 2011.

Meng CY, Smith BL, Bodhidatta L, Richard SA, Vansith K, Thy B et al. Etiology of diarrhea in young children and patterns of antibiotic resistance in Cambodia. Pediatr Infect Dis J. 2011;30(4):331-5. doi: 10.1097/INF.0b013e3181fb6f82.

Ministry of Agriculture and Rural Development, Viet Nam. Second National Strategy and Action Plan for Disaster Mitigation and Management in Viet Nam - 2001 to 2020. Hanoi; 2001.

Ministry of Environment, Cambodia. Cambodia's Initial National Communication. Phnom Penh; 2002.

Ministry of Environment, Cambodia and UNDP Cambodia. Building Resilience: The future for rural livelihoods in the face of climate change: Cambodia Human Development Report. 2011. Phnom Penh; 2011.

Ministry of Environment and Green Development, Mongolia. Climate Change Coordination Office (CCCO) (http://climatechange.gov.mn/index.php/en/).

Ministry of Health and Welfare, Republic of Korea. The development of climate change impact monitoring system and adaptation strategies for human health, 2008. Seoul; 2008.

Ministry of Health, Cambodia. Climate Change Strategic Plan for Public Health. Phnom Penh; 2012.

Ministry of Health, Cambodia. Health Strategic Plan 2008–2015. Phnom Penh; 2008.

Ministry of Health, Cambodia and World Health Organization. Climate Change and Health in Cambodia: a vulnerability and adaptation assessment. Phnom Penh: Ministry of Health, Cambodia; 2010.

Ministry of Health, Lao People's Democratic Republic. Climate Change and Health Adaptation Strategy (CCHAS) in Lao People's Democratic Republic. Vientiane; 2011:1–62 p.

Ministry of Health, Viet Nam. Action plan: response to the global climate change period 2010–2015. Ha Noi; 2010.

Ministry of Natural Resources and Environment, Viet Nam. Viet Nam Initial National Communication. Ha Noi; 2003.

Ministry of Natural Resources and Environment, Viet Nam. Viet Nam's Second National Communication to the UNFCCC. Ha Noi: 2010.

Ministry of Nature and the Environment, Mongolia. Mongolia's Initial National Communication. Ulaanbaatar; 2001.

Ministry of Nature, Environment and Tourism, Mongolia, United Nations Environment Programme and United Nations Development Programme. Mongolia: Assessment Report on Climate Change 2009. Ulaanbaatar: Ministry of Nature, Environment and Tourism; 2009.

Ministry of Nature, Environment and Tourism, Mongolia. Market Mechanism Country Fact Sheet: Mongolia. Ulaanbaatar; 2010.

Ministry of Planning, Kingdom of Cambodia. Migration in Cambodia: Report of the Cambodian Rural Urban Migration Project. Phnom Penh: 2012.

Moss RH, Babiker M, Brinkman S, Calvo E, Carter T, Edmonds J et al. Towards new scenarios for analysis of emissions, climate change, impacts, and response strategies. Geneva: Intergovernmental Panel on Climate Change; 2008.

Moss RH, Edmonds JA, Hibbard KA, Manning MR, Rose SK, van Vuuren DP et al. The next generation of scenarios for climate change research and assessment. Nature. 2010;463:747–756. doi: 10.1038/nature08823.

Muth S, Sayasone S, Odermatt-Biays S, Phompida S, Duong S, Odermatt P. Schistosoma mekongi in Cambodia and Lao People's Democratic Republic. Adv Parasitol. 2010;72:179-203. doi: 10.1016/S0065-308X(10)72007-8.

National Department of Health. National Action Plan for Climate Change and Health in Papua New Guinea. Port Moresby.; 2011.

National Disaster Centre, Papua New Guinea. Emergency and Disaster Management and Disaster Risk Reduction in Papua New Guinea. Port Moresby; 2005:1–25 p.

National Institutes of Health. Climate change vulnerability and impact assessment tools, M&E framework and compendium of adaptation practices in the Philippine health sector. Manila; 2011 (http://climate.gov.ph/images/events/Health1.pdf).

National Oceanic and Atmospheric Administration. Article first published online: 20 September 2012. DOI: 10.1029/97JC02906 (http://www.pmel.noaa.gov/pubs/outstand/mcph1720/TMP938460841.htm).

National Oceanic and Atmospheric Administration. Global Monitoring Division. CO2 at NOAA's Mauna Loa Observatory reaches new milestone: tops 400 ppm. 2013b (http://www.esrl.noaa.gov/gmd/news/7074.html accessed at 15 Oct 2013).

National Oceanic and Atmospheric Administration. National Centers for Environmental Information. State of the Climate: Global Analysis for Annual 2012 [published online January 2012] (http://www.ncdc.noaa.gov/sotc/global/2012/13).

National Oceanic and Atmospheric Administration. NOAA's El Niño Portal. (http://www.elnino.noaa.gov/, accessed 22 May 2013).

Nuorteva P, Keskinen M, Varis O. Water, livelihoods and climate change adaptation in the Tonle Sap Lake area, Cambodia: learning from the past to understand the future. Journal of Water and Climate Change. 2010;1(1):87-101. doi:10.2166/wcc.2010.010.

Office of Climate Change and Development, Government of Papua New Guinea. Interim action plan for climate-compatible development. Port Moresby; 2010.

Office of Climate Change and Development, Government of Papua New Guinea. Office of Climate Change and Development [website]. 2013 (http://www.occd.gov.pg/, accessed 29 May 2013).

Onozuka D, Hashizume M. Weather variability and paediatric infectious gastroenteritis. Epidemiol Infect. 2011b;139(9):1369–78 (http://www.ncbi.nlm.nih.gov/pubmed/21044404, accessed 1 Dec 2013).

Pacific Climate Change Science Program. Current and future climate of Papua New Guinea. 2011.

Pagnarith Y, Kumar V, Thaipadungpanit J, Wuthiekanun V, Amornchai P, Sin L, Day NP, Peacock SJ. Emergence of pediatric melioidosis in Siem Reap, Cambodia. Am J Trop Med Hyg. 2010;82(6):1106-12. doi: 10.4269/ajtmh.2010.10-0030.

Patz JA, Olson SH. Climate change and health: global to local influences on disease risk. Ann Trop Med Parasitol. 2006;100(5-6):535–49.

Pfeiffer M, Dora C. Health impact assessment in development lending – a reference guide and discussion draft. Geneva: World Health Organization; 2009.

Pham HV, Doan HT, Phan TT, Minh NN. Ecological factors associated with dengue fever in a Central Highlands province, Vietnam. BMC Infect Dis. 2011;11:172. doi: 10.1186/1471-2334-11-172.

Philippines National Department of Health. Health Promotion [website]. 2013 (http://www.doh.gov.ph/health_promotion_materials.html, accessed 10 June 2013).

Pinto E, Coelho M, Oliver L, Massad E. The influence of climate variables on dengue in Singapore. Int J Environ Health Res. 2011;21(6):415-26.

Pruss-Ustun A, Corvalan C. Preventing Disease through Healthy Environments. Towards an estimate of the environmental burden of disease. Geneva: World Health Organization; 2006.

Public Health Information and Geographic Information Systems (GIS). World Health Organization; 2008.

Quigley R, den Broder L, Furu P, Bond A, Cave B, Bos R. Health Impact Assessment International Best Practice Principles. Fargo: International Association for Impact Assessment; 2006.

Rammaert B, Beauté J, Borand L, Hem S, Buchy P, Goyet S et al. Pulmonary melioidosis in Cambodia: a prospective study. BMC Infect Dis. 2011;11:126. doi: 10.1186/1471-2334-11-126.

Regional Climate Change Adaptation Knowledge Platform for Asia. Scoping Assessment on Climate Change Adaptation in the Philippines. 2012.

Research Center for Rural Population and Health Viet Nam. Setting up a database on climate change in health sector in Viet Nam. Ha Noi; 2011.

Revich BA, Shaposhnikov DA. Extreme temperature episodes and mortality in Yakutsk, East Siberia. Rural Remote Health. 2010;10(2):1338.

Roebbel N. Health in the green economy: Health co-benefits of climate change mitigation – housing sector. Geneva: World Health Organization; 2011.

Royal Danish Embassy Development Cooperation Section. The Atlas of Cambodia: Data Tool Box. 2013 (http://www.cambodiaatlas.com/, accessed 24 May 2013).

Saji NH, Goswami BN, Vinayachandran PN, Yamagata T. A dipole mode in the tropical Indian Ocean. Nature. 1999;401(6751):360-3.

Sengchandala S. Climate Change Policy of Lao People's Democratic Republic – Objective of this Presentation. July 2010.

Seng H, Sok T, Tangkanakul W, Petkanchanapong W, Kositanont U, Sareth H, Hor B, Jiraphongsa C. Leptospirosis in Takeo Province, Kingdom of Cambodia, 2003. J Med Assoc Thai. 2007;90(3):546-51.

Socialist Republic of Viet Nam. National Strategy for Environmental Protection until 2010 and vision toward 2020. Hanoi; 2003.

Srivastava S. Coupling Multi-hazard Early Warning System and Emergency Communications in Mongolia. 2011.

United Nations Development Programme. Mongolia Human Development Report 2011. From Vulnerability to Sustainability: Environment and Human Development. Ulaanbaatar; 2011.

United Nations Development Programme. Promoting Social Equality in the Gobi-Areas of South Mongolia by Fostering Human Security with Integrated and Prevention Approaches. Proposal for the UN Trust Fund for Human Security. 2010.

United Nations Development Programme. UNDP Climate Change Country Profiles Vietnam. 2008 (http://ncsp.undp.org/sites/default/files/Vietnam.oxford.report.pdf).

United Nations Environment Programme. Viet Nam and climate change: a discussion paper on policies for sustainable human development. Ha Noi; 2009.

United Nations Framework Convention on Climate Change. GHG data – UNFCCC [online database] (http://mdgs.un.org/unsd/mdg/SeriesDetail.aspx?srid=749 and http://unfccc.int/ghg_data/ghg_data_unfccc/items/4146.php, accessed 12 April 2013).

United Nations Viet Nam. Climate Change Fact Sheet: greenhouse gas emissions and options for mitigation in Viet Nam, and the UN's responses. 2013.

Van NH, Trang DTH. Safe hospitals in emergencies and hospitals' resilience to climate change. Ha Noi; 2011.

Victoriano AF, Smythe LD, Gloriani-Barzaga N, Cavinta LL, Kasai T, Limpakarnjanarat K et al. Leptospirosis in the Asia Pacific region. BMC Infect Dis. 2009;9:147. doi: 10.1186/1471-2334-9-147.

Vlieghe E, Kruy L, De Smet B, Kham C, Veng CH, Phe T et al. Melioidosis, Phnom Penh, Cambodia. Emerg Infect Dis. 2011;17(7):1289-92. doi: 10.3201/eid1707.101069.

Warrilow D, Northill JA, Pyke AT. Sources of dengue viruses imported into Queensland, Australia, 2002–2010. Emerg Infect Dis. 2012;18(11):1850–7.

Watershed Resource and Environment Administration Government of Lao People's Democratic Republic. Strategy on Climate Change of the Lao People's Democratic Republic. 2010.

Wijedoru LP, Kumar V, Chanpheaktra N, Chheng K, Smits HL, Pastoor R et al. Typhoid fever among hospitalized febrile children in Siem Reap, Cambodia. J Trop Pediatr. 2012;58(1):68-70. doi: 10.1093/tropej/fmr032. Epub Apr 20, 2011.

World Bank. Climate risk and adaptation country profiles: Cambodia. 2013. (http://sdwebx.worldbank.org/climateportalb/home.cfm?page=country_profile&CCode=KHM, accessed 4 April 2013).

The World Bank Group. Economics of adaptation to climate change – Vietnam. 2010.

The World Bank Group. Vulnerability, risk reduction, and adaptation to climate change. Lao People's Democratic Republic; 2011a.

The World Bank Group. Vulnerability, risk reduction, and adaptation to climate change. Papua New Guinea; 2011b.

The World Bank Group. Vulnerability, risk reduction, and adaptation to climate change. Vietnam, 2011c. World Health Organization European Centre for Health Policy. Health impact assessment: main concepts and suggested approach. Gothenburg Consensus Paper. Gothenburg: WHO Regional Office for Europe; 1999.

World Health Organization. Workshop report: climate variability and change and their health effects in Pacific island countries. Samoa, 2000.

World Health Organization. Climate change and human health – risks and responses summary. Geneva; 2004.

World Health Organization. Using climate to predict infectious disease epidemics. Geneva: World Health Organization; 2005.

World Health Organization. Climate change and health: agenda item of the sixty-first World Health Assembly. 2008a.

World Health Organization. Climate change and health: report by the Secretariat. 2008b;1–6 p.

World Health Organization. Protecting health from the effects of climate change: Resolution of the Regional Committee for the Western Pacific. 2008c.

World Health Organization. Protecting health from climate change: connecting science, policy and people. Geneva, 2009a.

World Health Organization. Climate change and health: Mongolia [unpublished report]. 2009b.

World Health Organization. Climate change and health in Papua New Guinea. Manila, 2010a.

World Health Organization. Regional Office for Europe. Health Systems Crisis Preparedness Assessment Tool. 2010b

World Health Organization. WHO Consultation on the Essential Public Health Package to Enhance Climate Change Resilience in Least Developed Countries. Geneva; 2010c.

World Health Organization. WHO Country Cooperation Strategy for Mongolia 2010–2015. Geneva; 2010d (http://www.who.int/countryfocus/cooperation_strategy/ccs_mng_en.pdf).

World Health Organization. Health in the green economy: health co-benefits of climate change mitigation – housing sector. Geneva; 2011.

World Health Organization. Health in the green economy: health co-benefits of climate change mitigation – transport sector. Geneva; 2011a (http://www.who.int/hia/examples/trspt_comms/hge_transport_lowresdurban_30_11_2011.pdf).

World Health Organization. Western Pacific Region Country Health Information Profiles. Geneva; 2011b.

World Health Organization. Global Strategy for Dengue Prevention and Control 2012–2020. Geneva; 2012a.

World Health Organization. Regional Office for the Western Pacific. Strengthen control of vectorborne diseases to lessen the impact of climate change in the Western Pacific Region with focus on Cambodia, Mongolia, and Papua New Guinea: final project report. Manila; 2012b.

World Health Organization. World Malaria Report 2012. Geneva; 2012c.

World Health Organization, Department of Health Philippines. Health service delivery profile Philippines. 2012.

World Health Organization, Ministry of Health, Lao People's Democratic Republic. Health service delivery profile: Lao People's Democratic Republic. 2012.

World Health Organization, Ministry of Health, Mongolia. Health service delivery profile: Mongolia. 2012.

World Health Organization, Ministry of Health, Viet Nam. Health service delivery profile: Viet Nam. 2012.

World Health Organization, World Meteorological Organization. Atlas of health and climate. Switzerland; 2012.

World Health Organization. Protecting health from climate change vulnerability and adaptation assessment. Geneva; 2013.

World Health Organization. Quantitative risk assessment of the effects of climate change on selected causes of death, 2030s and 2050s. Geneva; 2014.

World Health Organization Regional Committee for the Western Pacific resolution WPR/RC59. R7 on protecting health from the effects of climate change. Manila: WHO Regional Office for the Western Pacific; 2008 (http://www2.wpro.who.int/rcm/en/archives/rc59/rc_resolutions/WPR_RC59_R7.htm).

World Health Organization Regional Office for Europe. Methods of assessing human health vulnerability and public health adaptation to climate change. Health and Global Environmental Change series No.1; 2003 (http://www.euro.who.int/__data/assets/pdf_file/0009/91098/E81923.pdf).

World Health Organization Regional Office for Europe. Tools for the monitoring of Parma Conference commitments: Meeting report. Copenhagen; 2010 (http://www.euro.who.int/__data/assets/pdf_file/0019/134380/e94788.pdf).

World Health Organization. Health in the Green Economy. 2013 (Retrieved from HIA: http://www.who.int/hia/green_economy/en/).

Wuthiekanun V, Pheaktra N, Putchhat H, Sin L, Sen B, Kumar V et al. Burkholderia pseudomallei antibodies in children, Cambodia. Emerg Infect Dis. 2008;14(2):301-3. doi: 10.3201/eid1402.070811.

Xu L, Liu Q, Stige LC, Ben Ari T, Fang X, Chan KS et al. Nonlinear effect of climate on plague during the third pandemic in China. Proc Natl Acad Sci USA. 2011;108(25):10214-9. doi: 10.1073/pnas.1019486108.

Xu Z, Etzel RA, Su H, Huang C, Guo Y, Tong S. Impact of ambient temperature on children's health: a systematic review. Environ Res. 2012;117:120–31.

Yang L, Chen PY, He JF, Chan KP, Ou CQ, Deng AP et al. Effect modification of environmental factors on influenza-associated mortality: a time-series study in two Chinese cities. BMC Infect Dis. 2011;11:342.

Young TK, Mäkinen TM. The health of Arctic populations: does cold matter? Am J Hum Biol. 2010;22(1):129-33. doi: 10.1002/ajhb.20968.

Yusuf AA, Francisco H. Climate change vulnerability mapping for Southeast Asia. Singapore: Economy and Environment Program for Southeast Asia; 2010.

Annex 1. A61/14

SIXTY-FIRST WORLD HEALTH ASSEMBLY — **A61/14**
Provisional agenda item 11.11 — **20 March 2008**

Climate change and health

Report by the Secretariat

1. There is now a strong, global scientific consensus that warming of the climate system is unequivocal,[10] and is caused by human activity, primarily the burning of fossil fuels which releases greenhouse gases into the atmosphere. Already, evidence from around the world shows that global warming is changing rainfall and storm patterns, and disrupting the balance of natural systems that supply the necessities of life.

2. WHO has, for several years, stressed that the health risks posed by climate change are significant, distributed throughout the globe, and difficult to reverse. Recent changes in climate have had diverse impacts on health, such as the death of more than 44 000 people during the heat wave in Europe in 2003. Climate-sensitive risk factors and illnesses are currently among the most important contributors to the global burden of disease; these include undernutrition (estimated to kill 3.7 million people per year), diarrhoea (1.9 million) and malaria (0.9 million). Such conditions and other health outcomes will be increasingly affected by accelerating climate change through its adverse effects on food production, water availability and the population dynamics of vectors and pathogens; already, for example, evidence shows that higher temperatures are increasing the risk of malaria transmission in the East African highlands.

Summary

3. Climate change will affect, in profoundly adverse ways, some of the most fundamental determinants of health: food, air and water. The warming of the planet will be gradual, but the increasing frequency and severity of extreme weather events, such as intense storms, heat waves,

10. Intergovernmental Panel on Climate Change, Fourth Assessment Report. Climate change 2007: synthesis report. Summary for policy-makers. Geneva, Intergovernmental Panel on Climate Change, November 2007 (unedited copy).

droughts and floods, will be abrupt and the consequences will be acutely felt. The earliest and most severe threats are to developing countries, with negative implications for the achievement of the health-related Millennium Development Goals and for health equity. It is therefore essential to formulate a clear response in order to protect human health and ensure that it is placed at the centre of the climate debate.

HEALTH ISSUES

4. The health sector, at international, national and subnational levels, has a responsibility, political leverage and staff with many of the necessary skills to protect the public from climate-related threats to health. Health professionals bring an understanding of primary prevention (analogous to strategies to mitigate climate change[11]) and secondary prevention (analogous to measures for adapting to climate change[12]) to the discussion of how to reduce and prevent climate-related disease, injury and death. Key concepts that should be considered in designing responses include the following.

5. **Climate change threatens public health security.** Global warming is expected to pose direct threats to health by causing more severe storms, floods, droughts and fires, with consequent disruptions in water and food supplies and medical and other services. Higher temperatures will change the distribution, and increase the burden, of various vector-borne, foodborne and water-related infectious diseases. The worsening of air quality, particularly owing to ozone pollution, increases the prevalence of asthma and respiratory infections, the number of admissions to hospital, and days of work and schooling lost. Meeting increasing energy demands by greater use of fossil fuels will tend to increase the number of cases of these air pollution-related illnesses and all-cause and all-age premature deaths. Greater frequency and intensity of heat waves will increase mortality and the incidence of heat stress and heat stroke. Evidence shows that this is already occurring.

6. **Health impacts will be disproportionately greater in vulnerable populations.** Globally, people at greatest risk include the very young, the elderly, and the medically infirm. Low-income countries and areas where undernutrition is widespread, education is poor, and infrastructures are weak will have most difficulty adapting to climate change and related health hazards. Vulnerability is also determined by geography, and is higher in areas with a high endemicity of climate-sensitive diseases, water stress, low food production and isolated populations. The populations considered to be at greatest risk are those living in small-island developing states, mountainous regions, water-stressed areas, mega cities and coastal areas in developing countries (particularly the large urban agglomerations in delta regions in Asia), and also poor people and those unprotected by health services. A major concern is the fact that some African countries have a high burden of climate sensitive diseases and poor public health capability to respond; the effects of climate change on socioeconomic development will seriously undermine health and well-being of people in such countries.

---

11. Mitigation in this context means action to reduce human effects on the climate system: principally strategies to reduce greenhouse gas emissions.
12. Adaptation in this context means adjustment in natural or human systems in response to actual or expected climatic stimuli or their effects, which moderates harm or exploits beneficial opportunities.

7. **Mitigating the effects of climate change can have direct and immediate health benefits.** A number of proposed mitigation strategies may improve health. For example, lessening the reliance on coal-fired generation of power will reduce air pollution, and associated respiratory and cardiopulmonary disease and death. Providing opportunities for the use of active transport (bicycling and walking) can also reduce levels of ambient air pollution, traffic-related injury and death, and obesity rates. Production and transport of food are major emitters of greenhouse gases.

8. **Adaptation is needed because some degree of climate change is inevitable, even if greenhouse gas emissions were abruptly capped. Failure to respond will be costly in terms of disease, health-care expenditure and lost productivity.** Estimated direct and indirect health-care costs and lost income due to several environmental illnesses (e.g. those caused by air pollution) often match or exceed the expenditure needed to tackle the environmental hazard itself.

ACTIONS

9. The overarching goals for the international response to protect health from climate change are: (a) to ensure that concerns about public health security are placed at the centre of the response to climate change; (b) to implement adaptive strategies at local, national and regional levels in order to minimize impacts of climate change on the health of human populations; and (c) to support strong actions to mitigate climate change and to avoid further dramatic and potentially disastrous impacts on health. These goals can be achieved by working through existing public health frameworks with the following specific objectives.

10. **Raise awareness of the need to ensure public health security by acting on climate change.** Strong, evidence-based and consistent advocacy by the global health community will be needed to raise awareness that global public health needs to be protected from climate change. Such awareness raising will call for health-sector professionals to show leadership in supporting rapid and comprehensive actions, promoting mitigation and adaptation strategies that both improve health now and reduce future impacts of climate change. The case for public health security should be made more clearly in national and international processes that guide policy and resources for work on climate change, such as preparation of National Communications and National Adaptation Programmes of Action, and the global Nairobi work programme on impacts, vulnerability and adaptation to climate change, under the United Nations Framework Convention on Climate Change. WHO can support this objective through its own advocacy within and outside the United Nations system, and by providing guidance to Member States' health sectors on how to engage more effectively in the above processes.

11. **Strengthen public health systems to cope with the threats posed by climate change.** Increased investment in public health systems is already necessary in order to meet the health-related Millennium Development Goals, whose achievement will be further compromised by the impact of climate change. For this reason, additional system strengthening and forward planning will be required. Within this broad context, at national level the health sector should: (a) assess the potential impacts of climate change on health; (b) review the extent to which existing health systems can cope with the additional threat posed by climate changes, and (c) develop and implement adaptation strategies to strengthen key functions that already protect against climatic risks. This approach will need to encompass interventions within the formal health sector, such as control of neglected tropical diseases and provision of primary health care, and actions to

improve the environmental and social determinants of health, from provision of clean water and sanitation, to enhancing the welfare of women. A common theme must be ensuring health equity and giving priority to protecting the health security of particularly vulnerable groups. WHO can provide technical support for building capacity to assess vulnerability and plan adaptive measures, and can mobilize and guide international support for the necessary strengthening of public health systems.

12. **Enhance capacity to deal with public health emergencies.** There is a particular need to strengthen systems to be able to respond to acute shocks associated with climate variability, including the health consequences of natural disasters, and more frequent, severe and wide-ranging epidemics. WHO can assist this effort through existing international programmes on health action in crises, and disease surveillance, reporting and response.

13. **Promote health development.** National and subnational health agencies can promote health through assessment of the health implications of decisions taken in other sectors, such as urban planning, transport, energy supply, food production, land use and water resources. In this way, they can support those decisions that provide opportunities for improving health and at the same time reduce emissions of greenhouse gases that cause climate change; these opportunities include new investment in sustainable transport in developed and rapidly developing countries and in clean domestic energy in developing nations. WHO's role could be to provide technical guidance and adapt tools (such as cost-benefit analysis and health impact assessment) for global and regional assessments of the implications for health of policies in sectors such as energy, transport and water and sanitation.

14. **Enhance applied research on health protection from climate change.** Better evidence is needed of the effectiveness and efficiency of public health measures to protect health from climate change. Such activities require systematic, interdisciplinary applied research in Member States. WHO can assist by working with research bodies throughout the world to define and promote a common research agenda, and facilitating information exchange among countries.

15. **Monitor and evaluate delivery.** National and subnational agencies should improve identification and monitoring of the health status of vulnerable groups, and evaluate the effectiveness of interventions aiming to protect health better from climate change. WHO can support this work through technical guidance in many areas including design of indicators, and working closely with existing international mechanisms for monitoring progress towards attainment of the health-related Millennium Development Goals.

16. **Foster cross-disciplinary partnerships.** In order to ensure wide-ranging and effective mitigation and adaptation, Member States should build partnerships at the national and subnational levels, exploiting the expertise of government agencies, intergovernmental and nongovernmental organizations, and community, industry and professional groups for health protection. WHO can support this process at national and international levels through further development of the multisector and cross-disciplinary "healthy settings" approach (e.g. healthy homes, schools, public spaces and work places).

17. An earlier version of this report was considered by the Executive Board at its 122nd session. The Board considered a draft resolution proposed by several Member States and adopted resolution EB122.R4.

ACTION BY THE HEALTH ASSEMBLY

18. The Health Assembly is invited to consider the draft resolution contained in resolution EB122.R4.

Annex 2. WHA61.19

SIXTY-FIRST WORLD HEALTH ASSEMBLY

WHA61.19

Agenda item 11.11

24 May 2008

Climate change and health

The Sixty-first World Health Assembly,

Having considered the report on climate change and health;[13]

Recalling resolution WHA51.29 on the protection of human health from risks related to climate change and stratospheric ozone depletion and acknowledging and welcoming the work carried out so far by WHO in pursuit of it;

Recognizing that, in the interim, the scientific evidence of the effect of the increase in atmospheric greenhouse gases, and of the potential consequences for human health, has considerably improved;

Noting with concern the recent findings of the Intergovernmental Panel on Climate Change that the effects of temperature increases on some aspects of human health are already being observed; that the net global effect of projected climate change on human health is expected to be negative, especially in developing countries, small island developing States and vulnerable local communities which have the least capacity to prepare for and adapt to such change, and that exposure to projected climate change could affect the health status of millions of people, through increases in malnutrition, in death, disease and injury due to extreme weather events, in the burden of diarrhoeal disease, in the frequency of cardiorespiratory diseases, and through altered distribution of some infectious disease vectors;

Noting further that climate change could jeopardize achievement of the Millennium Development Goals, including the health-related Goals, and undermine the efforts of the Secretariat and Member States to improve public health and reduce health inequalities globally;

Recognizing the importance of addressing in a timely fashion the health impacts resulting from climate change due to the cumulative effects of emissions of greenhouse gases, and further recognizing that solutions to the health impacts of climate change should be seen as a joint

13. Document A61/14 (This document is attached in Annex 2).

responsibility of all States and that developed countries should assist developing countries in this regard;

Recognizing the need to assist Member States in assessing the implications of climate change for health and health systems in their country, in identifying appropriate and comprehensive strategies and measures for addressing these implications, in building capacity in the health sector to do so and in working with government and nongovernmental partners to raise awareness of the health impacts of climate change in their country and take action to address them;

Further recognizing that strengthening health systems to enable them to deal with both gradual changes and sudden shocks is a fundamental priority in terms of addressing the direct and indirect effects of climate change for health,

1. REQUESTS the Director-General:

(1) to continue to draw to the attention of the public and policy-makers the serious risk of climate change to global health and to the achievement of the health-related Millennium Development Goals, and to work with FAO, WMO, UNDP, UNEP, the United Nations Framework Convention on Climate Change secretariat, and other appropriate organizations of the United Nations, in the context of United Nations reform initiatives, and with national and international agencies, to ensure that these health impacts and their resource implications are understood and can be taken into account in further developing national and international responses to climate change;

(2) to engage actively in the UNFCCC Nairobi Work Programme on Impacts, Vulnerability and Adaptation to Climate Change, in order to ensure its relevance to the health sector, and to keep Member States informed about the work programme in order to facilitate their participation in it as appropriate and access to the benefits of its outputs;

(3) to work on promoting consideration of the health impacts of climate change by the relevant United Nations bodies in order to help developing countries to address the health impacts of climate change;

(4) to continue close cooperation with Member States and appropriate United Nations organizations, other agencies and funding bodies in order to develop capacity to assess the risks from climate change for human health and to implement effective response measures, by promoting further research and pilot projects in this area, including work on:

(a) health vulnerability to climate change and the scale and nature thereof;

(b) health protection strategies and measures relating to climate change and their effectiveness, including cost-effectiveness;

(c) the health impacts of potential adaptation and mitigation measures in other sectors such as marine life, water resources, land use, and transport, in particular where these could have positive benefits for health protection;

(d) decision-support and other tools, such as surveillance and monitoring, for assessing vulnerability and health impacts and targeting measures appropriately;

(e) assessment of the likely financial costs and other resources necessary for health protection from climate change;

(5) to consult Member States on the preparation of a workplan for scaling up WHO's technical support to Member States for assessing and addressing the implications of climate change for health and health systems, including practical tools and methodologies and mechanisms for facilitating exchange of information and best practice and coordination between Member States, and to present a draft workplan to the Executive Board at its 124th session;

2. URGES Member States:

(1) to develop health measures and integrate them into plans for adaptation to climate change as appropriate;

(2) to build the capacity of public health leaders to be proactive in providing technical guidance on health issues, be competent in developing and implementing strategies for addressing the effects of, and adapting to, climate change, and show leadership in supporting the necessary rapid and comprehensive action;

(3) to strengthen the capacity of health systems for monitoring and minimizing the public health impacts of climate change through adequate preventive measures, preparedness, timely response and effective management of natural disasters;

(4) to promote effective engagement of the health sector and its collaboration with all related sectors, agencies and key partners at national and global levels in order to reduce the current and projected health risks from climate change;

(5) to express commitment to meeting the challenges posed to human health by climate change, and to provide clear directions for planning actions and investments at the national level in order to address the health effects of climate changes.

Eighth plenary meeting, 24 May 2008 A61/VR/8

Annex 3. WPR/RC59.R7

WORLD HEALTH ORGANIZATION — ORGANISATION MONDIALE DE LA SANTE

RESOLUTION

REGIONAL COMMITTEE FOR THE WESTERN PACIFIC — COMITE REGIONAL DU PACIFIQUE OCCIDENTAL

WPR/RC59.R7
26 September 2008

PROTECTING HEALTH FROM THE EFFECTS OF CLIMATE CHANGE

The Regional Committee,

Recalling resolution WHA51.29 on the protection of human health from risks related to climate change and stratospheric ozone depletion, resolution WHA61.19 on climate change and health, and resolution WPR/RC56.R7 on environmental health, all of which call for action to reduce the health impact of climate change;

Recognizing with concern the recent findings of the Intergovernmental Panel on Climate Change that the effects of temperature increases on some aspects of human health are already being observed, and that the net global effect of projected climate change on human health is expected to be negative, especially in developing countries, small island developing states and vulnerable local communities;

Noting with concern that the regional consultations held in 2007 identified current and emerging climate change-related health risks in the Region to include heat stress and waterborne and foodborne diseases associated with extreme weather events, vectorborne diseases, respiratory diseases due to air pollution and aeroallergens, food and water insecurity, malnutrition and psychosocial impacts from displacement;

Mindful that there are ongoing efforts to improve health systems to combat these health impacts, but climate change may require additional efforts to strengthen adaptive capacity of health systems to climate change;

Acknowledging that capacity to assess and minimize the health risks of climate change is limited, particularly in developing countries, especially Pacific island countries and areas;

Noting that WHO will present a draft workplan to the Executive Board at its 124th session on addressing the implications of climate change for health and health systems;

Recognizing the need for the health sector to advocate for decisions on mitigation and adaptation to climate change by other sectors, which will protect and promote health at the same time, and participate in the national and international processes that guide policy and resources for work on climate change;

Having reviewed the draft Regional Framework for Action to Protect Human Health from the Effects of Climate Change in the Asia Pacific Region,

1. ENDORSES the Regional Framework for Action to Protect Human Health from the Effects of Climate Change in the Asia Pacific Region as a guide for planning and implementing actions to protect health from the effects of climate change, while noting that a global workplan is expected to be adopted at the Sixty-second World Health Assembly to ensure coherence across WHO regions, a strong evidence base and non-duplication of efforts;

2. URGES Member States:

(1) to develop national strategies and plans to incorporate current and projected climate change risks into health policies, plans and programmes to control climate-sensitive health risks and outcomes;

(2) to strengthen existing health infrastructure and human resources, as well as surveillance, early warning, and communication and response systems for climate-sensitive risks and diseases;

(3) to establish programmes to reduce greenhouse gas emissions by the health sector;

(4) to assess the health implications of the decisions made on climate change by other sectors, such as urban planning, transport, energy supply, food production and water resources, and advocate for decisions that provide opportunities for improving health;

(5) to facilitate the health sector to actively participate in the preparation of national communications and national adaptation programmes of action;

(6) to actively participate in the preparation of a workplan for scaling up WHO's technical support to Member States for assessing and addressing the implications of climate change for health;

3. REQUESTS the Regional Director:

(1) to provide technical guidance and support to Member States for health vulnerability and adaptation assessment and to collaborate with other relevant organizations and Member States to undertake studies of the health impact of climate change;

(2) to strengthen country-level support to build national capacities to develop and implement national strategies and plans on mitigation and adaptation to climate change;

(3) to support Member States with training programmes on methodologies in the assessment and management of health risks due to climate change;

(4) to provide technical guidance on best practices of adaptation and mitigation within the health sector;

(5) to collaborate region-wide and with centres on mechanisms to share pertinent information, provide technical expertise for capacity-building and, taking into account global mandates, monitor the implementation of the Regional Framework for Action to Protect Human Health from the Effects of Climate Change in the Asia Pacific Region;

(6) to enhance cooperation with United Nations organizations and programmes, bilateral development assistance agencies, the private sector and development banks to increase the resources to implement the Regional Framework for Action;

(7) to report periodically to the Regional Committee on the progress made in implementing the Regional Framework for Action.

Eighth meeting, 26 September 2008

WPR/RC59/SR/8

Annex 4.

REGIONAL FRAMEWORK FOR ACTION TO PROTECT HUMAN HEALTH FROM EFFECTS OF CLIMATE CHANGE IN THE ASIA-PACIFIC REGION

Preamble

During the last 100 years, human activities related to the burning of fossil fuels, deforestation and agriculture have led to a 35% increase in the carbon dioxide (CO_2) levels in the atmosphere, causing increased trapping of heat and warming of the earth's atmosphere. The Fourth Assessment Report (AR-4) of the Intergovernmental Panel on Climate Change (IPCC) states that most of the observed increase in the globally-averaged temperatures since the mid-20th century was very likely due to the increase in anthropogenic greenhouse gas (GHG) concentrations. Eleven of the last 12 years (1995-2006) rank among the 12 warmest years in the instrumental record of global surface temperature. The IPCC also reports that the global average sea level rose at an average rate of 1.8 mm per year from 1961 to 2003. The total rise in the sea level during the 20th century was estimated to be 0.17 m.

The globally averaged surface warming projected for the end of the 21st century (2090–2099) will vary between 1.1–6.4 degrees centigrade. The global mean sea level is projected to rise by 30–60 cm by the year 2100, mainly due to thermal expansion of the ocean. It is very likely that hot extremes, heat waves and heavy precipitation events will continue to become more frequent. It is likely that future tropical cyclones (typhoons and hurricanes) will become more intense, with larger peak wind speeds and heavier precipitation, causing loss of life and an increase in injuries. These climatic changes will cause disruption of the ecosystem's services to support human health and livelihood, and will impact health systems. The IPCC projects an increase in malnutrition and consequent disorders, with implications for child growth and development. The disruption in rainfall patterns can be expected to lead to an increased burden of diarrhoeal disease and to the altered spatial distribution of some infectious-disease vectors. WHO estimates that the modest anthropogenic climate change that has occurred since 1970, claims 150 000 lives annually.

Therefore, the IPCC urges a drastic reduction in GHG emissions to mitigate global warming and an urgent implementation of adaptation measures.

The current and emerging climate change-related health risks in Asia and the Pacific include heat stress and water- and food-borne diseases (e.g. cholera and other diarrhoeal diseases) associated with extreme weather events (e.g. heat waves, storms, floods and flash floods, and droughts); vector-borne diseases (e.g. dengue and malaria); respiratory diseases due to air pollution; aeroallergens, food and water security issues; malnutrition; and psychosocial concerns from displacement. These risks and diseases are not new, and the health sector is already tackling these problems. However, the capacity to cope with potentially increasing levels of these risks and diseases is limited, particularly in developing countries.

There is a growing, but still limited, political commitment to integrate health considerations into efforts to mitigate and adapt to climate change at national and international levels in the

Region. Also, there is also insufficient awareness among the general public about climate change and its impact on health.

The availability of relevant hydro-meteorological, socioeconomic and health data is limited and available data are often inconsistent and seldom shared in an open and transparent manner. Furthermore, there is insufficient capacity for assessment, research and communication on climate-sensitive health risks in many countries, as well as insufficient capacity to design and implement mitigation and adaptation programmes. There is an urgent need to incorporate health concerns into the decisions and actions of other sectors while they plan to mitigate and adapt to climate change, to ensure that these decisions and actions also enhance health. By promoting the use of non-motorized transport systems (e.g. bicycles) and fewer private vehicles, greenhouse gas emissions would be reduced, air quality would improve and more people would be physically active. Such an approach would produce associated benefits (i.e. reduce the burden of disease while lowering greenhouse gas emissions) and needs to be promoted.

Goal and objectives of the regional framework

Goal:

To build capacity and strengthen health systems in countries and at the regional level to protect human health from current and projected risks due to climate change.

Objectives:

(1) Increase awareness of health consequences of climate change;

(2) strengthen the capacity of health systems to provide protection from climate-related risks and substantially reduce health system's greenhouse gas emissions; and

(3) ensure that health concerns are addressed in decisions to reduce risks from climate change in other key sectors.

Recommended actions

Objective 1: To increase awareness of health consequences of climate change

Governments, through relevant agencies, should:

(1) Undertake studies on the health implications of climate change and share information to understand how to promote changes in individual and corporate behaviours that mitigate climate-related health risks, while protecting and promoting health.

(2) Enhance political commitment and strengthen institutional capacity and arrangements to achieve adaptation and mitigation goals.

(3) Facilitate national working groups, nongovernmental organizations and civil society to develop coordinated mitigation and adaptation plans by including relevant sectors, regions and disciplines.

(4) Develop awareness-raising programmes and learning resource materials to educate and engage a broad range of stakeholders, including local communities, health and other relevant professionals, and the media on the potential health impacts of climate variability and change and on appropriate measures to reduce climate-sensitive risk factors and adverse health outcomes.

WHO should:

(1) Provide specific climate change-related technical guidance for vulnerability and adaptation assessments and surveillance systems, which provide methods for identifying risks to vulnerable groups, quantifying the burden of disease from climate change, and quantifying costs and benefits of health adaptation measures to ensure comparability across countries.

(2) Support countries in the development of vulnerability and adaptation assessment and analysis tools, and in the development of a set of indicators on climate change-related health risks.

(3) Encourage and facilitate regional knowledge-sharing and networking on climate change and human health within the health sector as well as between disciplines.

Objective 2: To strengthen health systems capacity to provide protection from climate-related risks, and substantially reduce health system's GHG emissions

Governments, through relevant agencies, should:

(1) Develop and implement national action plans for health that are integrated into existing national plans on adaptation and mitigation to climate change.

(2) Develop integrated strategies to incorporate current and projected climate change risks into existing health policies, plans and programmes to control climate-sensitive health outcomes, including integrated vector management, and health risk management of disasters.

(3) Strengthen existing infrastructure and interventions, including human resource capacity, particularly surveillance, monitoring and response systems and risk communication, to reduce the burden of climate-sensitive health outcomes. Key concerns vary by country; common concerns include vector borne diseases, air quality and food and water security.

(4) Strengthen public health systems and disaster/emergency preparedness and response activities, including psychosocial support, through increased collaboration and cooperation across sectors. This should include documentation, sharing and evaluation of the effectiveness of local knowledge and practices.

(5) Provide early warning systems to support prompt and effective responses to current and projected health burdens. In order to achieve this, national and regional climate forecasting information, including climate change projections, should be fully utilized.

(6) Implement adaptations over the short, medium or long term; be specific to local health determinants and outcomes of concern; and facilitate the development of community-based resource management. The costs and benefits of different interventions should be determined.

(7) Establish climate change focal points or mechanisms within national health institutions to ensure the implementation, monitoring and evaluation of health mitigation and adaptation actions and ensure that health issues are adequately addressed in these actions.

(8) Establish programmes through which then health sector substantially reduces GHG emissions; by doing so, it could also serve as a best practice model for other sectors.

WHO should:

(1) Facilitate greater contribution of funds from donor agencies for climate change- and health-related programme implementation.

(2) Support countries technically and financially to build national capacities to develop and implement national action plans on mitigation and adaptation, including conducting research on the health impacts of climate change.

(3) Support countries technically and financially by providing training programmes on methodologies and assisting in the assessment and management of health risks due to climate change.

(4) Develop and provide technical guidance on good adaptation and GHG emission reduction practices within the health sector.

Objective 3: To ensure that health concerns are addressed in decisions to reduce risks from climate change in other key sectors^.

Governments, through relevant agencies, should:

(1) Develop integrated strategies to incorporate current and projected climate change risks into existing policies, legislation, strategies and measures of key development sectors to control climate-sensitive health outcomes. Examples include the promotion of public and non-motorized transportation, clean energy and disaster risk management.

(2) Facilitate the health sector to actively participate in national communications to the United Framework Convention on Climate Change (UNFCCC), and include health issues as the core elements in the negotiation process.

(3) Ensure active health participation in the national climate change team.

WHO should:

(1) Support the establishment of a regional centre on climate change and health, which has links to results of vulnerability and adaptation assessments and data sources, both between countries within and outside the Region, and links to hydro-meteorological services at global, regional and national levels. This centre will support a regional network of practitioners working on climate change and health, with access to international technical expertise to facilitate the sharing of best practices.

(2) Identify and establish WHO collaborating centres on climate change and health in the Region.

Annex 5. EB136/16

EXECUTIVE BOARD
136th session
Provisional agenda item 7.2

EB136/16
5 December 2014

Health and the environment

Climate and health: outcome of the WHO Conference on Health and Climate

Report by the Secretariat

1. This report covers two topics: the outcome of the WHO Conference on Health and Climate (Geneva, 27–29 August 2014) and a revised WHO work plan on climate change and health.

2. The WHO Conference on Health and Climate marked a major step in responding to the requests of the Sixty-first World Health Assembly to the Director-General in resolution WHA61.19, adopted in 2008.

THE WHO CONFERENCE ON HEALTH AND CLIMATE

3. The overall objective of the Conference was to provide the health and sustainable-development communities with the most up-to-date and authoritative evidence, tools and information in order: to enhance population resilience to, and protect health from, climate change; to identify the health benefits associated with reducing emissions of greenhouse gases and other climate pollutants; and to support health-promoting policies on climate change.

4. The Conference was further intended to contribute the health perspective to the United Nations Climate Summit 2014 (New York, 23 September 2014), and to reinforce health ministers'

participation in national and international policy discussions in preparation for the Conferences of the Parties to the United Nations Framework Convention on Climate Change to be held in Lima in December 2014 and Paris in December 2015.

5. The Conference was attended by some 400 participants, including 25 ministers, from 96 Member States in all WHO regions. The heads of four United Nations entities (WHO, WMO, the secretariat of the United Nations Framework Convention on Climate Change, and the secretariat of the United Nations Office for Disaster Risk Reduction) also participated together with representatives of civil society organizations, experts and health practitioners.

6. In order to set the example of the health community reducing its own environmental impact, WHO applied for the first time the United Nations' guidance on "green meetings". The Secretariat minimized the printing of documents, made maximum use of electronic documents and webcasting, and provided vegetarian and locally-sourced food that had minimal associated greenhouse gas emissions. The Conference was also the first carbon-neutral WHO meeting, with the greenhouse gas emissions associated with the travel of all participants offset through the purchase of carbon credits through the secretariat of the United Nations Framework Convention on Climate Change.

EVIDENCE PRESENTED AND CONCLUSIONS OF THE CONFERENCE[1]

7. Evidence was presented that human actions, principally the burning of fossil fuels and associated release of climate pollutants, are causing significant changes to the global climate system. At the current pace of emissions of greenhouse gases, average surface temperatures are expected to rise by 4 °C by the year 2100.[2] Conservative estimates suggest that climate change will cause some 250 000 additional deaths per year before the middle of the current century.[3] The main risks to health are expected to be more intense heatwaves and fires; increased prevalence of food-, water- and vector-borne diseases; increased likelihood of undernutrition resulting from diminished food production in poor regions; and lost work capacity and reduced labour productivity in vulnerable populations.

8. Less conclusive but still concerning evidence exists for other risks, including: breakdown in food systems and increased prevalence of violent conflict associated with resource scarcity and population movements; exacerbation of poverty stemming from a slow-down in economic growth, with negative implications for achieving health targets including those of the Millennium Development Goals and the objectives of the post-2015 sustainable development agenda currently under discussion. Poorer populations and children are disproportionately

---

1 A full report of the Conference by the International Institute for Sustainable Development's Reporting Services has been published on the WHO website; see http://www.who.int/globalchange/mediacentre/events/climate-health-conference/en/ (accessed 18 November 2014).

2 IPCC. Summary for policymakers. In: Stocker TF, Qin D, Plattner G-K, Tignor M, Allen SK, Boschung J et al, Eds. Climate change 2013: the physical science basis contribution of Working Group I to the Fifth Assessment Report of the Intergovernmental Panel on Climate Change. Cambridge, England, and New York: Cambridge University Press; 2013.

3 WHO. Quantitative risk assessment of the effects of climate change on selected causes of death, 2030s and 2050s. Geneva: World Health Organization; 2014.

4 WHO. Air pollution estimates: summary of results and method descriptions. Geneva: World Health Organization; 2014. See also the accompanying document on air pollution and health, EB136/15

5. United Nations. Climate, health jobs: thematic discussion at Climate Summit 2014. http://www.un.org/climatechange/summit/2014/08/climate-health-jobs/ (accessed 6 October 2014).

at risk of the effects of climate change, with different impacts on women and men. Overall, the impact is likely to widen existing health inequalities, both between and within populations.

9. Protection of health against climate change risks can be enhanced through ensuring better and more equitable access to services that mitigate and improve the social and environmental determinants of health, strengthening of basic public health interventions, and interventions targeted at climate- related risks.

10. The opportunity exists for policies that reduce the extent of climate change to yield also significant, local, near-term health benefits, in particular by reducing the annual mortality attributable to household and ambient air pollution (about 4.3 million and 3.7 million, respectively).[4]

11. Health can be improved by greener and more sustainable choices in various sectors, including household energy, electricity generation, transport, urban planning and land use, buildings, food and agriculture. For example, both the greater use of renewables in electricity generation and more efficient combustion of fossil fuels and biomass can cut ambient air pollution. Putting such policies into practice can translate into significant health cost-savings, particularly through reductions in the burden of noncommunicable diseases.

12. The health sector can also improve its own practices and at the same time minimize its carbon emissions. Health services in developed countries are major consumers of energy and significant emitters of greenhouse gases; energy efficiency, shifting to renewables, and greener procurement and delivery chains can both improve services and cut carbon emissions. In contrast, many health facilities in the poorest countries lack any electricity supply; for resource-constrained settings and off-grid hospitals and clinics, low-carbon energy solutions can form an important component of an overall energy supply strategy.

13. The Conference underlined the importance of meeting the challenges in line with the mandates from the Health Assembly, the United Nations Framework Convention on Climate Change and related processes, making use of existing mechanisms and building on the rapidly emerging experience worldwide. It also recognized the willingness of WHO to host a platform to further develop coordinated efforts on health and climate change with its expanding range of partners now active in this field.

14. The main messages and outcomes were presented at the United Nations Climate Summit 2014.[5] WHO is working with the United Nations Framework Convention on Climate Change secretariat and the governments of France and Peru to promote health on the agendas of the forthcoming Conferences of the Parties to the Convention.

---

[1] See document WHA67/2014/REC/3, summary record of twelfth meeting of Committee A of the Sixty-seventh World Health Assembly, section 9H, and document WHA66/2013/REC/3, summary record of seventh meeting of Committee B of the Sixty-sixth World Health Assembly, section 1E.

[2] See document EB124/2009/REC/1, Annex 1.

THE WHO WORK PLAN ON CLIMATE CHANGE AND HEALTH

15. In January 2009, the Executive Board in resolution EB124.R5 endorsed the proposed WHO work plan on climate change and health. In response to subsequent requests of Member States,[1] the Secretariat has revised the existing work plan.[2] The central focus is on environmental determinants of health (one of WHO's leadership priorities for 2014–2019), but several actions require consideration of climate change within other strategic priorities of the Organization.

Work plan aims

16. The revised work plan will provide support to Member States: to respond to the health risks presented by climate change, by strengthening the resilience of health systems to climate risks and improving their capacity to adapt to long-term climate changes; and to identify, assess and promote actions that reduce the burden of diseases associated with air pollution, and other health consequences of policies that also cause climate change.

17. WHO will implement the work plan with a particular focus on promoting health equity. It will take into account variation in the vulnerability of populations to climate risks, and in their capacity to respond, both of which are associated with gender and other social determinants of health. The work plan will focus in particular on improving the health of the most vulnerable population groups, including the poor, children and the elderly. Its implementation will be in line with the Twelfth General Programme of Work 2014-2019.

18. The main proposed changes in the proposed work plan are (i) establishment of a partnership "platform" to respond to the increasing number of activities and actors engaged in this field; (ii) greater emphasis on actions that can improve health while also mitigating the extent of climate change; and (iii) promoting the need and providing tools for more systematic provision of country- specific information and monitoring of progress.

Objective 1. Strengthen partnerships to support health and climate within and outside the United Nations system

19. *Action 1.1* Establish a stable partnership platform to enable WHO to work with other organizations that have complementary capacities (for example, nongovernmental organizations on awareness raising, collaborating centres on research, and development banks on financing). This action will support and build on existing partnerships on specific issues, such as the Global Framework for Climate Services and the Climate and Clean Air Coalition to reduce Short-Lived Climate Pollutants, and develop new partnerships for specific thematic areas, such as the linkages between climate change, health, water and sanitation, and nutrition.

20. *Action 1.2* Continue to provide leadership on health throughout the system-wide response of the United Nations to climate change. Particular attention will be given to ensuring that health is appropriately reflected in policy and planning processes, and financial support mechanisms under the United Nations Framework Convention on Climate Change, the Post-2015 Framework for Disaster Risk Reduction and the post-2015 sustainable development goals currently being discussed.

Objective 2. Awareness raising

21. *Action 2.1* Develop tools, guidance, information and training packages for raising awareness of the links between health and climate, and the potential for enhancing health through

mitigation of the extent of climate change. The focus will be on supporting national health decision-makers to engage effectively on setting policies for adaptation and mitigation, for example in negotiations being carried out under the United Nations Framework Convention on Climate Change.

22. *Action 2.2* Further develop WHO's networks and mechanisms for disseminating information to the wider community of health professionals, and the general public, working particularly with relevant nongovernmental organizations and youth groups.

Objective 3. Promote and guide the generation of scientific evidence

23. *Action 3.1* Monitor and guide research agendas. This action will include the formulation of mechanisms to support exchange between researchers and decision-makers, definition of regional and national research agendas on climate change and health, and monitoring the extent to which research output is responsive to the priorities identified by the Health Assembly in resolution WHA61.19.

24. *Action 3.2* Lead, or contribute to, international assessments of the risks to health from climate change and of the benefits to health of mitigation policies, in collaboration with partners, including the Intergovernmental Panel on Climate Change.

25. *Action 3.3* Further develop and support the use of tools for Member States to assess the effectiveness of interventions to increase resilience to climate change, and the health impacts of adaptation and mitigation decisions in other sectors. These evaluations should include assessment of economic consequences and wider sustainable-development implications, including the cost–effectiveness of interventions; the costs of inaction; and health benefits from mitigation and adaptation policies.

26. *Action 3.4* Produce and systematically maintain country-specific profiles, including hazards, vulnerabilities and projected impacts, as well as the potential for health gains from health-promoting interventions to increase resilience and mitigate the extent of climate change.

Objective 4. Provide policy and technical support to the implementation of the public health response to climate change

27. *Action 4.1* Support a more systematic approach to increasing the resilience of public health systems to climate, by providing an operational framework that identifies health functions that should be strengthened and that should take account of climate risks. This action would encompass public health interventions within the formal health sector, and cross-sectoral action to improve the environmental and social determinants of health, ranging from improved air quality and wider access to clean water and sanitation to enhanced disaster preparedness.

28. *Action 4.2* Support capacity-building through the setting of norms and standards, development of technical guidance, and training courses. This action will include key areas such as the use of information on climate to improve disease surveillance and early warning; enhanced health preparedness for and response to extreme weather events; and opportunities to simultaneously tackle climate change and air pollution.

29. *Action 4.3* Implement pilot projects to test new approaches. These activities will be supported by a clear strategy for expansion and mainstreaming into policies, plans and programmes of health and health-determining sectors, such as water and sanitation, agriculture and energy provision.

30. *Action 4.4* Provide specific policy and technical support on health facilities, including: ensuring resilience to climate change risks; provision of environmental services, including access to electricity, clean water and sanitation, and waste management; and reduction of greenhouse gas emissions from health sector operations.

31. *Action 4.5* Provide guidance and technical support to Member States for accessing financial resources to enhance health protection from climate change risks, and valuation of health benefits in cross-sectoral policies. The Secretariat will serve as a clearing house for funding opportunities, and provide support through advocacy, monitoring and dissemination of evidence in order to ensure appropriate access and share of resources for health.

32. *Action 4.6* Establish a voluntary system for countries to report their progress in increasing the resilience of health systems to climate change and gaining health benefits from mitigation policies, using an agreed set of indicators. This action will also provide a systematic and objective basis for reporting to WHO's governing bodies.

IMPLEMENTATION OF THE REVISED WORK PLAN

33. The revised work plan will be implemented through WHO's programmes at all levels and partnerships supported by the platform referred to in Action 1.1. Impact will be monitored and evaluated with the indicators referred to in Action 4.6 through progress reports to the Health Assembly. Monitoring and evaluation will also take place through the mechanisms and indicators in the Twelfth Programme of Work, 2014–2019. The Secretariat's in this area has expanded significantly since the previous work plan was endorsed, but within the same budget for the strategic priority. The programme budgets for future bienniums will need to be reassessed on the basis of the requirements of Member States for collaboration and support.

ACTION BY THE EXECUTIVE BOARD

34. The Board is invited to note the report and provide guidance on the revised work plan.

= = =